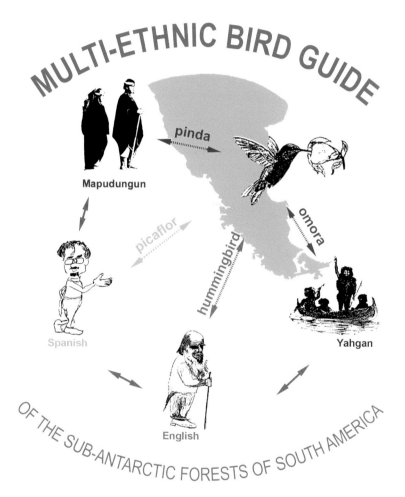

MULTI-ETHNIC BIRD GUIDE

Mapudungun

pinda

omora

picaflor

hummingbird

Spanish

English

Yahgan

OF THE SUB-ANTARCTIC FORESTS OF SOUTH AMERICA

MULTI-ETHNIC BIRD GUIDE
OF THE SUB-ANTARCTIC FORESTS OF SOUTH AMERICA

Texts and general direction
Ricardo Rozzi

&

Ethnographic research and story collection
Francisca Massardo & Ricardo Rozzi

Ornithological research
**Ricardo Rozzi, Francisca Massardo, Christopher B. Anderson,
George Clark, Guillermo Egli, Steven McGehee & Eduardo Ramilo**

Yahgan bird names and stories
Úrsula Calderón & Cristina Calderón

Mapuche bird names and stories
Lorenzo Aillapan

Biographies
Cristina Zárraga & Lorenzo Aillapan

Narrators of bird names and stories recorded on the CDs
Yahgan: **Úrsula Calderón & Cristina Calderón**
Mapudungun: **Lorenzo Aillapan**
Spanish: **Francisca Massardo, Luis Gómez & Lorenzo Aillapan**
English: **Kurt Heidinger, Uta Berghöefer, Steven McGehee &
Lorena Peñaranda**

Recording and sound production
John Schwenk, Nolberto González & Ricardo Rozzi

SUB-ANTARCTIC BIOCULTURAL CONSERVATION PROGRAM
University of North Texas - Universidad de Magallanes

Library of Congress Cataloging-in-Publication Data

Guía Multi-étnica de aves de los bosques subantárticos de Sudamérica austral. English
Multi-ethnic bird guide of the sub-Antarctic forests of South America / Ricardo Rozzi, editor.
-- 2nd ed.
    p. cm.
Previously published in English under title: Multi-ethnic bird guide of the austral temperate forests of South America.
    In English, with bird names given also in Yahgan, Mapuche and Spanish.
    Includes bibliographical references and index.
    ISBN 978-1-57441-282-6 (cloth : alk. paper)
    1. Forest birds--Chile--Names. 2. Forest birds--Argentina--Names. 3. Forest birds--Chile--Identification.
    4. Forest birds--Argentina--Identification. 5. Yahgan Indians--Folklore. 6. Mapuche Indians--Folklore.
    7. Birdsongs--Chile. 8. Birdsongs--Argentina. I. Rozzi, Ricardo, 1960- II. Title. III. Title: Multiethnic bird guide of the Sub-Antarctic forests of South America.
    QL689.C5 G85 2010
    598.0983--dc22

                                                            2009052684

Universidad
de Magallanes

CONTENTS

Figure 2. Mapuche poet Lorenzo Aillapan (center), recording Mapuche bird stories in the forests of the Omora Ethnobotanical Park, Navarino Island with Steven McGehee (right) and Ricardo Rozzi (left).

and teachers, both Chilean and international, with whom we traveled to Navarino Island, Cape Horn County, in January 2000.

Living together with the Indigenous Yahgan Community of Bahía Mejillones and encountering the wisdom of the Yahgan grandmothers Úrsula Calderón and Cristina Calderón, permitted us to come closer to traditional Yahgan ornithological knowledge. During the first year of work on Navarino Island, the memories of the grandmothers Úrsula and Cristina, concerning the Yahgan bird names and narratives, initially were flowering rapidly and then later more gradually. I will never forget the moment when we had given the project up as finished and put away the recording equipment, when Grandmother Úrsula knocked on the door saying, "I remembered: *wichoa*, that is the Yahgan name of the red-breasted one" (Figure 3). That was the last name registered in this guide book.

Figure 3. The Yahgan grandmother Úrsula Calderón during a recording session with Ricardo Rozzi in Mejillones Bay, Navarino Island.

Figure 4.The recordings of the bird names and narratives were done in the forests of Mejillones Bay and the Omora Ethnobotanical Park on Navarino Island, between January 2000 and May 2001. In the photograph (from left to right), plant physiologist Francisca Massardo, Yahgan grandmother Úrsula Calderón, ornithologist Steven McGehee, Mapuche poet Lorenzo Aillapan, Yahgan grandmother Cristina Calderón, and teacher Luis Gómez (grandson of Grandmother Cristina) in Mejillones Bay, February 2001.

Figure 5. Members of the Yahgan community and ethnographers Martin Gusinde (at the left in the front row) and Wilhelm Koppers (at the right in the front row) who participated in the Chiexaus ceremony celebrated in Mejillones Bay in 1922. Gusinde and Koppers recorded Yahgan ecological knowledge, which constitute an essential reference for the analysis and understanding of the bird stories recorded by us eighty years later (Courtesy Anthropos Institute).

In the CD recordings that accompany this book only the names of the birds that were remembered by the grandmothers Úrsula and Cristina are recorded. Other Yahgan names appear written at the beginning of the texts for each species. With respect to the narratives, we adopted a different approach. After recording all of the stories that were remembered by the grandmothers, we decided together with teacher Luis Gómez (grandson of Grandmother Cristina) "to retell" to Úrsula and Cristina four Yahgan bird stories that possess a great educational value (Figure 4). These stories about the Magellanic Woodpecker, Andean Tapaculo, Buff-Necked Ibis and Greenback Firecrown Hummingbird had been told by their father Juan Calderón and other members of the Yahgan community to Austrian ethnographer Martin Gusinde between 1920 and 1923 (Figure 5). Such "intervention" had as its objective the stimulation of their memories and to compose this guide book as educational material that would contribute to the continuity of the Yahgan language and traditional ecological knowledge. In the case of the stories of the ibis, or *lejuwa*, and the woodpecker, or *lana*, the Grandmothers recovered a more or less living memory. In the case of the hummingbird, or *omora*, they only recalled the name of *omora* as was told by Grandmother Julia. The story of the tapaculo, or *tuto*, did not evoke any remembrance, and this story was recorded by Luis Gómez with his Grandmother Cristina based on the text registered by Martin Gusinde.

The first phase of work making the audio recordings culminated in February 2001, when the Mapuche poet Lorenzo Aillapan accepted our invitation to come to Navarino Island. During the summer, with the Grandmothers Ursula and Cristina, Lorenzo, and the whole team we recorded the names and stories of the birds in Yahgan, Mapudungun, Spanish and English, seated around the fire or under the canopy of the austral forests.

Ecologist Christopher Anderson prepared the English version of the Mapuche stories, working directly with the poet Lorenzo Aillapan and of the Yahgan stories with the plant physiologist Francisca Massardo. The texts were then edited by writer Kurt Heidinger, who also recorded many of the stories in English (Figure 6). Other stories and names in English were recorded by US ornithologist Steven McGehee, German geographer, Uta Berghöfer, and Chilean lawyer, Lorena Peñaranda, which allowed the incorporation of variations in tone and pronunciation in this language, which acquired importance at the austral extreme of the Americas with the arrival of the Anglican missionaries 150 years ago.

Figure 6. Writer Kurt Heidinger (left) and sound engineer John Schwenk (right) record and edit the sound files in the studios of WHUS Radio (University of Connecticut, USA) to compose the counterpoint between the diverse languages of birds and humans recorded in the CDs included in this book.

19

The Spanish version of the Mapuche stories was recorded by the poet Aillapan with their onomatopoeic imitations. The Yahgan stories were recorded in Spanish first by the grandmothers Úrsula and Cristina Calderón. However, in April 2001 poet and writer Cristina Zárraga, who is the granddaughter of Grandmother Cristina, arrived on Navarino Island. Cristina Zárraga agreed to collaborate and revise, together with her grandmother, the Spanish texts of the Yahgan narratives. Then, the majority of the Yahgan stories were recorded again by Cristina Zárraga in Spanish (Figure 7).

Figure 7. Writer Cristina Zárraga (left) and researcher Uta Berghöfer (center-left) record Yahgan stories with Ricardo Rozzi (center-right) in Spanish and English, while photographer Oliver Vogel (right) chronicles the scene on the north coast of Navarino Island.

The meeting with Cristina Zárraga gave origin to a fertile relationship that is expressed in her own work as a writer and educator, as well as an investigator of Yahgan culture, which is continued to this day. With the poet Aillapan, the meeting also has permitted the continuation of joint work about the avifauna. Another fruit of this shared work was publishing the CDs and book *Twenty Winged Poems of the Native Forests of Chile*, which combined the poems of Lorenzo with the recordings of songs and illustrations prepared by a team of ornithologists (Figure 8). Today these poems continue their linguistic drift in the composition of twenty musical pieces by Andrés Alcalde and other Chilean composers. As such, following the oral tradition of the Yahgan and Mapudungun languages, this guide book has been a dynamic process that has given rise to new outlooks and songs in the dialogue with birds at the tip of South America.

Figure 8. Nolberto González (left), music teacher in Puerto Williams and director of Button Records, and Ricardo Rozzi (right) edit the sound files of birds and human voices, for this book and Twenty Winged Poems *with Lorenzo Aillapan.*

II. The Biocultural Mosaic of Southwestern South America

In order to facilitate the understanding of the relationship between birds, cultures and types of forested habitats, here we present a brief synthesis of the mosaic of forest ecosystems and cultures in southwestern South America. The ethnographic and forest-type maps will serve as a reference when viewing the distribution maps that appear together with each bird species (Figures 9 and 10, see pages 22 and 23 respectively). Combining these maps it is possible to deduce in what types of forests and in which Mapuche, Yahgan or other indigenous territory each bird inhabits.

II.1 Biological diversity

The temperate forests of southwestern South America are the world's southernmost forested ecosystems (Figure 11). They extend for approximately 2,000 km from central Chile (~35°S) until the tip of the continent (56°S), and for a short distance along the eastern slope of the Andes Mountain Range in Argentina. In order to emphasize the high latitude of the long, narrow strip of southern Chilean and Argentinean temperate forests, we entitled this book the *Multi-Ethnic Bird Guide of the Sub-Antarctic Forests of South America*.

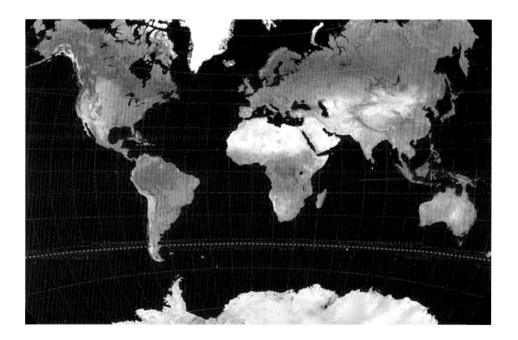

Figure 11. *World image showing that the sub-Antarctic Magellanic rainforest ecoregion hosts the southernmost forests of the world, extending nine degrees beyond the latitude of Stewart Island, New Zealand (47°S, indicated by the dashed line). This remote ecoregion is one the 24 remaining wilderness areas in the planet at the beginning of the 21st century.*

Forest types map

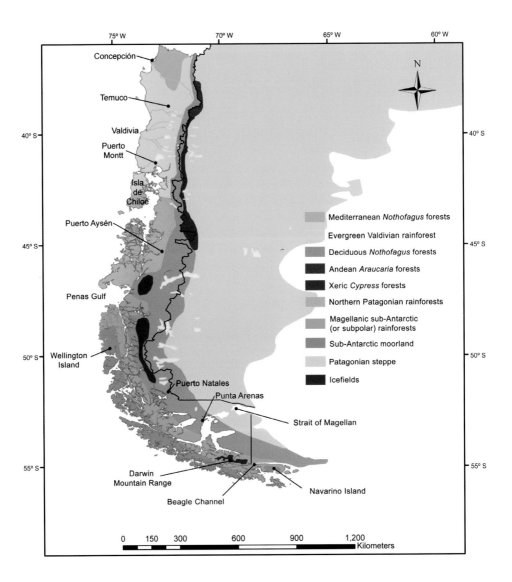

Figure 9. Map of the main forest types in the region of the southern South American temperate forests. The number and name of forest types vary with different classification systems, but there is consensus about their marked diversity in the region (Modified from Veblen et al. 1983, 1996, Gajardo 1994, Donoso 1995).

Ethnographic map

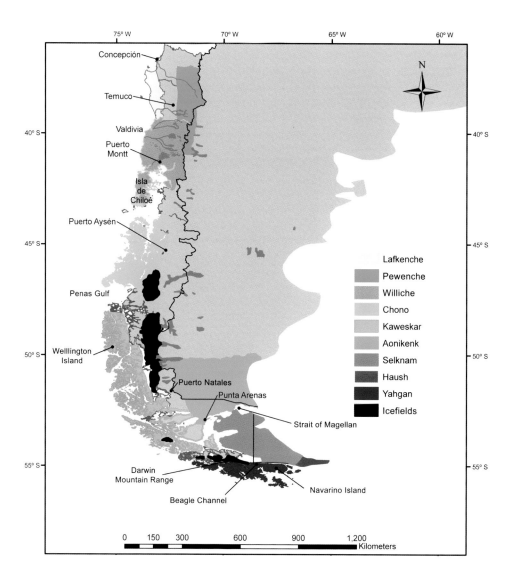

Figure 10. Ethnographic map of the temperate forest region of southern South America. (Modified from Cooper 1946, Hidalgo et al. 1996, McEwan et al. 1997).

The temperate forests that grow along the southwestern margin of South America include at its austral end the sub-Antarctic Magellanic evergreen rainforest ecoregion, which spans the myriad of archipelagoes from the Gulf of Penas (47ºS) to Cape Horn (56ºS); i.e., to almost 10 degrees south of all other forest ecosystems in the Southern Hemisphere (Figure 11). Due to its remote location, archipelagic geography, and lack of terrestrial connectivity and development, the temperate forest region, especially the sub-Antarctic Magellanic evergreen rainforest ecoregion, has remained relatively free from direct, modern human impact. Today, this region still maintains more than 70% of its original vegetation cover, over an area larger than 10,000 km², and the archipelagic zone (south of 42ºS) harbors one of the lowest human population densities within temperate latitudes (0.14 inhabitants/ km²). The South American temperate forest ecosystems, especially in the southern sub-Antarctic Magellanic rainforest ecoregion are embedded in a matrix of peatlands, bogs, and cushion bogs known as the Magellanic moorland complex. The area covered by the Magellanic moorland totals 4.4 million ha and ranks among the world's largest wetland ecosystems. For these reasons it has been identified as one of the 24 wilderness areas remaining in the planet.

Another determining characteristic of the southern forests' biological diversity is their biogeographical isolation, compared to other forest biomes. Vast geomorphological and climatic barriers separate the southern temperate forests from the nearest tropical forests, by at least 1,500 km. To the north lies the Atacama Desert, one of the driest deserts in the world. Westward and southward we find the Pacific Ocean. To the east, rise the high Andean Mountains and beyond them lies the vast, dry Patagonian steppe and other xeric plant communities in Argentina. In this way, the austral temperate forests of South America really are a biogeographic island. This insular character has promoted the evolution of a largely endemic biota. Close to 90% of the woody plant species are endemic. Among vertebrates, levels of endemism reach 80% for amphibians, 36% for reptiles, 33% for mammals and 30% for terrestrial birds. The latter percentage rises to 66%, if we consider only the southern extreme of South America. Hence, two-thirds of the bird species that you could observe in the region live only in southern Chile and Argentina, and half of these (that is, a third of all birds) are endemic to the forest habitats of the extreme austral region. Such a high degree of endemism in the avifauna is comparable to the levels found on oceanic islands, rather than the lower levels of endemism recorded in other forested continental regions of the world.

The fauna and flora of the South American temperate forests are also unique for three additional reasons. First, they possess an interesting variety of biogeographic origins, which include the contrasting biota of the Neotropical (warm) and the Gondwanic (sub-Antarctic, cool temperate) regions. Second, along the extensive latitudinal strip of forests, marked temperature and rainfall variations occur. Third, the South American temperate forest biome has a complex topography, including the Coastal Mountain Range, the Central Valley, the Andes, and the southern Archipelago zone, which generate further environmental heterogeneity. The combined effect of biogeographic, climatic, and topographic variations results in a remarkably diverse mosaic of forest ecosystems along the length of the latitudinal gradient (Figure 11).The forest type mosaic includes landscapes as diverse as the high Andean zones, dominated by the Pewen or the Monkey Puzzle Tree (*Araucaria araucana*), the dense and structurally complex Valdivian rainforests, the Andean-Patagonian deciduous forests, and the sub-Antarctic (or subpolar) Magellanic rainforests (Figures 12-15).

Figure 12 (next page). Forests dominated by Monkey Puzzle Trees or Pewen (Araucaria araucana) in the area of the high Bio-Bio River (38ºS).

Figure 13. Evergreen rainforests of the Valdivian ecoregion, with a complex canopy structure (left) and understory (below). Photographs were taken in the Alerce Andino National Park (41.5ºS).

Figure 14. *Deciduous forests dominated by Tall Deciduous Beech (*Nothofagus pumilio*) at the southern end of its distributional range. Photograph was taken in the Omora Ethnobotanical Park, Navarino Island (55ºS) in the Cape Horn Biosphere Reserve.*

Figure 15. *Sub-Antarctic forests on Hoste Island in the Cape Horn Biosphere Reserve. Low-lying areas are dominated by evergreen forests of Evergreen Beech (*Nothofagus betuloides*). In contrast, in higher zones deciduous forests dominated by Low Beech (*Nothofagus antarctica*) prevail. The latter are characterized by their bright red colors in autumn.*

27

II.2 Cultural diversity

The South American temperate forest biome hosts not only unique ecosystems and biota, but also a mosaic of idiosyncratic and diversified indigenous cultures. Among these Amerindian cultures we find the world's most southerly pre-Columbian inhabitants, the Yahgans. They occupy the archipelago territory of the extreme south of the Americas (55-56°S), located from the southern coast of Tierra del Fuego to Cape Horn (Figure 10). The Yahgans possess a long canoeing tradition, and they host a refined ornithological knowledge and ecological understanding of the sub-Antarctic ecosystems in the austral archipelago.

North to the Yahgan territory, Tierra del Fuego was inhabited by the Selknam or Ona and by the Haush. These were terrestrial hunter groups. North of the Strait of Magellan, the Patagonian steppe, which extends to the eastern border of the Andean-Patagonian forests, corresponds to the territory of another tribe of land hunters, the Aonikenk or Tehuelches. Along the western strip of these latitudes, the rainy region of the sub-Antarctic forests from Isla Wellington (49°S) to southwest Tierra del Fuego (54°S) are the territory of the Kaweskar or Alacaluf canoe culture (Figure 10). These southern Amerindian peoples host unique ecological knowledge and practices, associated with their capacity to inhabit areas with very rigorous climatic conditions. Their cultural singularity is due also to the degree of isolation that they experienced with respect to contact with other dominant South American cultures, such as the pre-Incan and Incan civilizations.

The northern half of the South American temperate forest region hosts the territories of the Mapuche people (35-42°S), whose name Mapuche means "People of the Land" (*Mapu* = land; *che* = people). This culture possesses a vital link with its environment and a capacity to communicate with the birds and other living beings. Indeed, the Mapuche speak *Mapudungun*, which means the "Language (= *dungu*) of the Land." The onomatopoeic character of the *Mapudungun* bird names illustrates the intimate relationship between their language and the land.

The vital link that the Mapuche people have with their land is further expressed in the names of their communities. Today, the three main Mapuche groups define themselves according to the specific habitat type they inhabit: i) in the Monkey Puzzle tree (*Araucaria araucana*) forests of the volcanic Andean mountain range of central-southern Chile and Argentina (Figure 9) live the *Pewenche* (Figure 10), i.e., people of the Monkey Puzzle tree (*pewen*), for whom the seeds of this tree play a central role in their subsistence and cultural habits; ii) in the coastal forests of central-southern Chile live the *Lafkenche,* i.e., people of the sea (= *lafken),* who are dependent on marine organisms (algae, mussels, fish) and coastal ecosystems, including farmlands; iii) in the evergreen rainforests of southern Chile live the *Williche*, i.e., people of the south (= *willi*), who are dependent on the plants and animals of the Valdivian forest ecoregion between the Toltén River (38°S) and the south of Chiloé Island (42°S). Formerly, the *Pikun Mapu*, the land of the north (= *pikun*), was inhabited by a fourth group of Mapuche, the *Pikunche* or people of the north. They lived in the Mediterranean sclerophyllous forests of Central Chile, where they conducted farming and husbandry activities, mainly between the Aconcagua River (32°S) and the Itata River (38°S). The *Pikunche* were gravely affected by the invasion of the Spaniard conquerors. Their culture has had a lasting influence on the toponymy (names of places) and the cultural traditions of Central Chile.

Today, the losses of cultural diversity are even greater than those of biological diversity. Half of the indigenous languages that were spoken in Chile are extinct, and a third of the remaining extant ones are seriously threatened. At the world-wide level, 90% of the 6,912 languages spoken today could cease to exist during the 21st century! In this context, this multi-ethnic book aspires to contribute to the continuity of the Yahgan and *Mapudungun* languages and cultures, as much as to the conservation of the bird songs and their ecological interactions with the temperate forests in southwestern South America.

III. Ethno-Ornithological Philosophy and Environmental Ethics

This guide is an invitation to wander through the forests of southwestern South America and listen to the voices of the birds and people that inhabit them. It provides an orientation to perceive the uniqueness of the ornithological voices and human expressions inhabiting each ravine, bay or slope in the world's southernmost forests.

Each time that we observe a bird, our perception is informed by something of the bird, and something of ourselves—our senses, instruments, and concepts. In this way, a dialogue arises between the birds themselves, and the names and stories that express our human perceptions of them. The variety of birds, with their distinct vocalizations and behaviors, is as great as the richness of ornithological names and narratives, with their particular human ways of looking at the birds and inhabiting the world. This book guides us to recognize the cultural and biological diversities, and to understand the dynamic bio-cultural relationships that take place in each of the Mapuche, Yahgan, and scientific bird stories.

Yahgan and *Mapudungun* are languages with an oral tradition, and within the forests it is more common to hear the birds, than it is to see them. Therefore, listening attentively helps to cultivate our intelligence and sensitivity in the identification and knowledge of birds, and the appreciation of indigenous names and stories. For this reason, we suggest beginning by listening to the recordings of the CDs that accompany this guidebook. Familiarize yourself with the birdcalls and their names in Yahgan, Mapudungun, Spanish and English. Each name, each sound, has a story.

Upon familiarizing yourself with the audio, recognizing the calls and names of the birds, you will be able to discover how the names and stories show us complementary facets of the birds, cultures, forested ecosystems, and their interactions in the austral extreme of the American Continent. For example, as illustrated in Figure 16, the species *Sephanoides sephaniodes* receives the names *omora*, *pinda*, *picaflor*, and *hummingbird*. The names in Mapudungun and English allude to the rapid, beating sound of the wings. *Pinda* or *pigda* refers to the sound that is produced when one thing is rubbed against another (=*pigudcun*). In English this peculiar sound produced by the beating of the wings of this *bird* is called *humming*, thus the name *hummingbird*. The names in Spanish and Yahgan allude, in contrast, to the behavioral aspects that can be observed in this bird. *Picaflor* denotes the habit of visiting "to pierce" (= *picar*) the flower (= *flor*) to drink its nectar. The valiant character possessed by *omora*, the tiny warrior in the Yahgan narratives, could be related to the territorial behavior of the hummingbird, who defends the flowers on which it feeds.

Figure 16. Sephanoides sephaniodes *is one of the endemic bird species of southwestern South America. The names given to this species by four of the languages spoken in this region illustrate the diversity of perspectives and worldviews about birds and their ecosystems, and ways of inhabiting them, and provide a basis for comparative analysis of indigenous and scientific ecological knowledge.*

In the face of the diversity of names and stories about each bird species, a question arises: How can so many narratives exist about the same bird? Are they all false or all true? Are some truer than others? In order to respond to these questions, this multi-ethnic guide book helps to interpret the traditional stories from the viewpoint of ecology, including the natural history and life history of the birds, as well as evolutionary and human ecology.

As a point of departure in the ecological outlook for listening and reading the ethno-ornithological stories, we propose the following analogy, shown in Figure 17. In the temperate forests of southern South America, the hummingbird (*Sephanoides sephaniodes*) and the bumblebee (*Bombus dahlbomii*) visit the flowers of the *kolkopiw* vine (*Philesia magellanica*). The hummingbird and the bumblebee take the nectar of this flower, obtaining the sugar and water that permits them to survive. On the other hand, upon visiting the flowers, the hummingbird and bumblebee help the reproduction of this plant by transporting pollen from flower to flower. In this way, more flowers will be produced and the cycle of life continues. This scene can be observed in the native forests of southern Chile and Argentina   -and it appears "natural." It would not occur to us that the bumblebee should teach the hummingbird to visit the *kolkopiw* flowers in the way of an insect—or vice—versa: the hummingbird should teach the bumble bee to visit the flowers in the way of a bird. Each species visits the flowers in its own manner and both survive in the austral ecosystems.

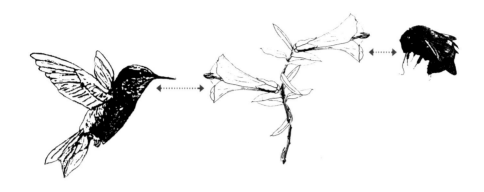

Figure 17. *The hummingbird* Sephanoides sephaniodes *and the bumblebee* Bombus dahlbomii *are two of the most common pollinators in the South American temperate forest biome. For example, both feed on the flower of the* koikopiw *(*Philesia magellanica*). However, while the* hummingbird *perceives the red color of the flowers, and feeds mainly on nectar sipped with its beak and tongue, the bumblebee sees the ultraviolet pattern of the same flower, and feeds mainly on pollen collected with its legs. In spite of these differences each one of these species has successfully co-evolved with the* koikopiw *flower, in its own way. This ecological and evolutionary understanding of inter-specific diversity might help to understand and respect the inter-cultural diversity in southwestern South America. The language of each culture is the result of an idiosyncratic co-evolutionary history with the birds and ecosystems, which today is expressed in the different names and narratives given by each of these cultures to each of the bird species included in this book.*

In an analogous way, we can interpret the diversity of Yahgan, Mapuche and Western-scientific viewpoints about birds. Each of these cultures, and languages, recognizes the bird species and interacts with them in a particular way. As has occurred with the hummingbird and bumblebee, the Mapuche and Yahgan cultures have inhabited and survived in the austral forests for thousands of years. The traditional ecological knowledge (including the recognition of the birds and their relationship with the biological communities and ecosystems) has been suitable to allow the survival of these peoples, with their own languages, and ecological practices. Therefore, their stories are not merely "myths" or "legends," but rather they interpret the natural reality in an effective manner.

At the beginning of the 21st century, the traditional ecological knowledge contained in the Yahgan and Mapuche ornithological narratives of this book invites us, as members of contemporary global society, to diversify our ways of knowing about, and inhabiting nature, and of living together with the birds and their ecosystems. This enriches our experience of visiting and exploring the austral forests of South America. At the same time, the broad array of traditional ecological knowledge offers us valuable perspectives regarding the conservation of these ecosystems and approaches to inhabiting them in a sustainable manner.

The Mapuche and Yahgan ornithological stories not only contrast scientific views, but also there are substantial similarities between them. For example, the indigenous narratives share two central notions with the contemporary, ecological-evolutionary perspective: 1) the sense of kinship between human beings and birds, derived from common genealogies or evolutionary histories, and 2) the sense of biotic communities or ecological networks, of which humans and birds form parts.

III.1 Our relatives, the birds

With respect to the evolutionary sense of kinship, modern biology has discovered that the human species (*Homo sapiens*) possesses cells, blood vessels, vertebrae and other structures that are very similar in birds. As can be observed on the cover of this book, the eyes of birds and human beings are very similar, as well. When walking through the forests, use your binoculars to discover how similar your own eyes are to the eyes of the birds. This type of similarity would be explained evolutionarily by the existence of a common origin or ancestor for birds and mammals. Such a common ancestor would have existed millions of years ago. In the broad sense of a common evolutionary history and biological nature, these affirmations that result from a scientific analysis share essential concepts with those contained in the Mapuche and Yahgan narratives, such as the sentence "in ancestral times when birds were humans" with which several Yahgan stories begin. Let's consider the following four examples (Figure 18): A) One of the Mapuche stories tells that Lautaro or *Leftraru*, the greatest of Mapuche warriors, descended from the lineage of eagles and caracaras or *traru* (*Caracara plancus*). So much so, that the name *Leftraru* means fast (=*lef*) *traru*. B) One of the Yahgan stories affirms that the woodpeckers, or *lana*, (*Campephilus magellanicus*) are descendants of a pair of Yahgan siblings. C) The tree of life by German biologist Ernst Haeckel illustrates the scientific-evolutionary theory of Charles Darwin, which proposes that human beings possess a common evolutionary origin shared with all living beings; however, in contrast to the previous Amerindian views, man is portrayed at the top of this tree. D) In order to illustrate equitable kinship relationships, on the cover of this *Multi-Ethnic Bird Guide of the Sub-Antarctic Forests of South America*, birds and humans are portrayed close to each other on branches of a High Deciduous Beech (*Nothofagus pumilio*) tree. On this species of tree, which is abundant in the Yahgan and Mapuche territories, each person is portrayed near a bird with which he or she has close affinities. For example, poet Lorenzo Aillapan (bottom right on the cover) turned into a Mapuche Bird-Man, *Uñümche*, after dreaming of a *Ñamku*, or Red Backed-Hawk, and receiving a feather, as he poetically says:

In the passing of the years, I maintain the prophetic dream.
In the dream the falling of a feather was a sign to me
of this hawk-bird. For this, I am the Mapuche Bird Man.
Since an early age, I have maintained a connection.

The genealogies of birds and humans illustrated in Figure 18 allude to an evolutionary history that implies a common nature between the bird and human species. From the point of view of contemporary environmental ethics, the three cultural perspectives—Mapuche, Yahgan and scientific—emphasize the *intrinsic value* of avifauna because *the birds are our distant evolutionary relatives*. This implies that, to some degree, the existence of birds can be subject to moral considerations based on ontological and ethical judgments on par with those we use to judge the value of human life.

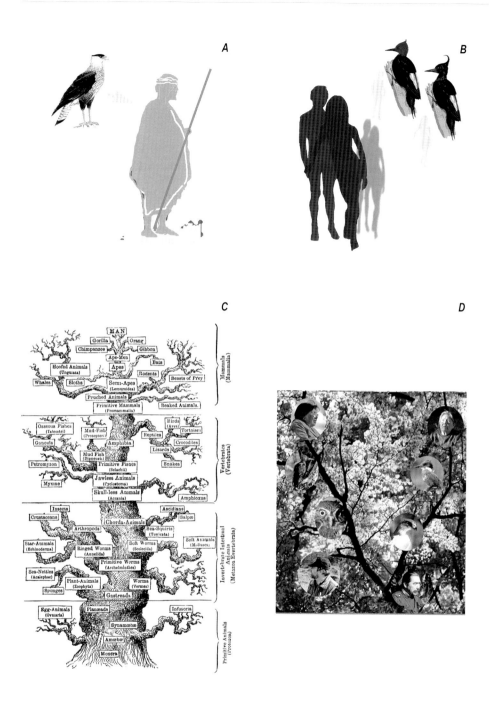

Figure 18. Common genealogies and kinship between birds and humans as depicted above by the illustrations of the indigenous Mapuche (A) and Yahgan (B) worldviews, as well as by the scientific perspective of the evolutionary tree of life by Ernst Haeckel (C), and by the design on the cover of this Multi-Ethnic Bird Guide.

III.2 Ecological and ethical networks

Regarding the concept of ecological networks, the participation of the human species in food webs and other biotic interactions is evident today for contemporary science. Human beings are not separate from nature; we constitute a keystone species for many ecosystems. We influence ecosystem processes, at the same time that ecosystems provide us with goods (such as medicinal plants, fishes and other food) and services (such as clean air, the regulation of climate and hydrologic cycles, increasing soil fertility, and the control of pests). The new discipline of ecological economics has determined, for example, that the cheapest and most sustainable method to assure the supply of drinking water is the protection of forested watersheds. Concepts such as these, which are recent for science, are rooted in the Mapuche and Yahgan traditional ecological knowledges. For example, the Yahgan story of the hummingbird, or *omora*, (*Sephanoides sephaniodes*), commands that water should be shared between all members of human and biotic communities, to maintain social and ecological well-being.

The perspective of conservation of bird and human communities in the Yahgan story of *omora* is comparable to contemporary ecological statements. Modern sciences propose that birds and other species contribute to ecological processes, such as pollination, the dispersal of seeds, the control of insect populations, or preying upon rodents. In the austral fjords (Figure 19), with their feces, or guano, or by carrying remains of marine prey, birds also play a key role in soil fertility and nutrient cycling by transporting minerals from ocean to mountain ecosystems. At the same time, the maintenance of these ecosystem processes contributes to the conservation of the flora in watersheds. This vegetation is crucial for the regulation of hydrologic flows, because it contributes to preventing floods as well as droughts. In this way, the Yahgan narrative of *omora*, as well as the scientific ecosystem analysis, recommends the protection of biodiversity of the watershed in order to secure the equitable and sustained provisioning of drinking water.

From the perspective of contemporary environmental ethics, the imperative implicit in the Yahgan narrative and the scientific analysis of "ecosystem services" is to conserve the bird community in order to assure reliable water. For this reason, we ought to respect the life of the diverse birds, humans and other living beings. This imperative emphasizes the notion of *instrumental value,* because the conservation of biodiversity is an *instrument* for the survival of humans, and all species.

The notions of *intrinsic* and *instrumental value* of the life of birds, humans and other living beings that co-inhabit the region of the temperate forests of South America, derive as much from current scientific analyses, as from traditional ecological knowledges. Its ethical-practical purpose is to promote the continuity of life, and the well-being of all co-inhabitants, humans non-humans.

Through a knowledge of the diverse languages of the cultures and the birds we can feel and look at ourselves immersed in the forests, living together with the birds and with the myriad of perceptions we humans can have of the birds. As human beings of different sexes, ages, geographies, cultures and histories, the austral forests open to us an opportunity—in the context of a globalized world—to glimpse and discover very diverse biocultural corners. This *Multi-Ethnic Guide* invites us to reflect upon how to live together respectfully with the singular biocultural diversity that inhabits the extreme south of the Americas. At the same time, its voices and narratives can inspire us to find respectful inter-specific and inter-cultural ways of co-habitation with the birds, the humans and ecosystems of our own urban, rural and other remote corners of the planet; each one characterized by its own unique biocultural diversities.

Figure 19. Close-up of the evergreen forests dominated by the Evergreen Beech (Nothofagus betuloides), which grow all the way to the high tide line along the coasts of fjords, channels and island that typify the Yahgan sub-Antarctic landscapes.

35

IV. The Specific Songs and Stories Recorded in this *Multi-Ethnic Bird Guide*

The song of each bird can vary from one moment to the next, between spring and winter, from individual to individual, between one forest and another—as occurs with the *chámuj*, *chinkol*, *chincol*, Rufous-Collared Sparrow or *Zonotrichia capensis*, whose trills vary between night and day and from northern to southern populations.

Rufous-Collared Sparrow (Zonotrichia capensis)

Like the voices of the birds, the names of and stories about birds told by the Yahgan, the Mapuche and the scientists change from moment to moment, from place to place and from person to person. For example, *omora* is more hummingbird-human in the story recorded by Grandmother Julia; in contrast, it is more human-hummingbird in the narrative told by Grandmother Rosa, but always *omora* is small and courageous. So, we find variations and similarities among stories, even within the narrations of a given story. The accounts of this guidebook constitute only a moment in the record of scientific, Yahgan and Mapuche ornithological stories. The diversity of bird and human voices remains endlessly open.

Grandmother Julia, who told Yahgan bird stories to Martin Gusinde in the 1920s (Courtesy Anthropos Institute).

Omora (Sephanoides sephaniodes)

Young Rosa at Mejillones Bay, who told Yahgan bird stories to Patricia Stambuk in the 1980s
(Photograph of young Rosa in 1917. Courtesy Museo Salesiano Maggiorino Borgatello).

This multi-ethnic bird guide does not represent the ethno-ornithological knowledge of native South American people in general. Instead, our intention is to present, as specifically as we can, the ways that two ethnic groups see the birds of the forests and co-inhabit with them. Furthermore, in this guidebook we find specific records of birdcalls and stories, which have many variations throughout the varied geographies, communities, and histories of the Yahgan and Mapuche territories in southwestern South America. If the stories depend on the story teller, then it is essential that we know about the people with whom we recorded the Yahgan and Mapuche bird names, and narratives included in the CDs of this book, through the biographies as told by themselves.

37

BRIEF BIOGRAPHIES OF

Úrsula Calderón and Cristina Calderón

Cristina Zárraga

Úrsula Calderón Harban was born on the 7th of September, 1925 in Mejillones Bay, Navarino Island. After the death of her mother, when she was barely 7 years old, she lived with her half-sister Dora and her brother Juan. When she was 14, she worked on the Róbalo Ranch, and at age 15, she married José González. Together, they worked hunting sea otters, and during eight years they lived moving from one place to another by boat. They resided in their tent in Puerto Navarino or simply where they anchored at night. Later, they set up residence on Mascar Island, but, as their children needed to attend school, Úrsula moved to Puerto Williams with the children. José remained on Mascar, raising cattle and sheep. Later, he too lived in Puerto Williams, where he died in 1987, brought down by lung cancer. Today, Úrsula lives with her three children in her house in Villa Ukika in Puerto Williams, and she dedicated herself to handcrafted artistry—weaving baskets of rush, building canoes, needlepoint and knitting—in addition to speaking her native language.*

Grandmother Cristina Calderón (left) and her sister Úrsula in the house at Mejillones Bay, Navarino Island, during a break while recordings were made for this multi-ethnic guidebook in February 2001.

Grandmother Úrsula Calderón died in January 2003. Today she rests in the Cemetery of Mejillones, Navarino Island.

*Grandmother Cristina Calderón weawing a basket made of rush or ushkulampi (*Marsippospermum grandiflorum*) in her craft working place in Villa Ukika, Navarino Island, June 2001.*

Cristina Calderón Harban was born on the 24th of May, 1928, in Robalo Bay, Navarino Island. Her infancy passed between the bays of Robalo and Mejillones. At a very early age her mother died in the latter place. As a result, she remained in the care of her grandfather, Alapainche, who in a short time also passed away. Then she lived with her cousin Clara and also her aunt Kerti. At the age of 16, she was required to live with Felipe Garay, a good man who, although quite a bit older than she (50 years), cared for her during the years that they lived together at Eugenia Ranch at the northern coast of Navarino Island, until the death of Felipe. From this union three children were born. At age 19, she met Luis Zárraga, a man descended from the disappeared Ona ethnic group. With him and the children, they lived at Harberton Ranch in Argentina for nine years, where six more children were born. Then, they moved to Puerto Williams on Navarino Island, building their house in Villa Ukika in 1960. Later, Luis became sick, dying later in Punta Arenas. In 1964, Cristina united with Teodocio González, with whom she shares her life until today. From this union one daughter was born. Cristina dedicates herself to handcrafted artistry—traditional weaving of rush baskets, making Evergreen Beech bark canoes, knitting and painting—in addition to fluently speaking the Yahgan language.

Villa Ukika, August 2001

SHORT BIOGRAPHY OF THE

Mapuche Bird-Man or *Üñümche*

Lorenzo Aillapan Cayuleo

He was born in the community of Rukatraru (habitat of the *traru*) on Lake Budi (*Furi leufü*), which lies back to back with the great Pacific Ocean on the southern cone of South America, and is the region's only saltwater lake. He was born with the name *Llankaleu* (precious stone or pearl of the water).

The poet Lorenzo Aillapan Cayuleo, Mapuche Bird-Man or Üñümche, with photographer Oliver Vogel (center), and ornithologist Steven McGehee (left) at the entrance of the Omora Ethnobotanical Park, Navarino Island, on February 27, 2001, after completing the recordings for this Multi-Ethnic Bird Guide.

He is the son of farming parents and the grandson of the powerful *lonko* (chief) Aillapan (nine lions or pumas). The *lonko* Aillapan is and was a warrior poet of the Mapuche style before and after the so-called "Pacification of the Araucanía" (when Chile finally took control of the Araucanian or Mapuche territory in about 1880).

The child Llankaleo (u) later will be named Lorenzo by the Civil Registry due to the demands of the Chilean law that require "saint's names" for people. At the age of 8 to 9 Llankaleu is consecrated and initiated through a dream (*pneuma*) as a bird-man (*üñümche*), bird spirit of the ancient Mapuche cosmovision. This consecration was guided by renowned masters: Master Huaiquián, Master Imio, Master/nuke Kayuleu (six rivers). The university of life with the guidance of wise masters gave him the title of *üñümche* (bird person) as an inheritance for his whole life; the community and all the people approved the title of Mapuche Bird-Man—until today.

At age 11, Lorenzo went for the first time in primary school (the Methodist School of Rucatraru), and he began to understand the Chilean/Spanish language. After completing his primary education, with the help of the North American benefactors Rann Crawford, Husell Sargent and Nancy Sargent, Lorenzo went to the Methodist Farming School of Nueva Imperial. He also studied at the Men's High School of Nueva Imperial. Later, he completed his higher education and worked in Santiago, the capital of Chile.

Traytrako, August 2001

EXPLANATION OF THE LAYOUT

Sequence of birds. This guidebook includes the names and narratives of fifty bird species. The sequence of birds follows an imaginary journey from the forest interior towards the margins and adjacent open habitats. The order of the bird species in the illustrated text is the same as the sequence of their recorded names in the tracks of the enclosed CD I. The sequence includes:

(1) *Birds that inhabit either exclusively or principally the forest interior (name tracks 2-9).* All of these species are year round residents, and are endemic to the temperate forests of southwestern South America; that is, they do not inhabit any other place on the planet. Most of these species are difficult to see because they live deep in the foliage or the understory of the forests; however, they can be detected and identified by their characteristic calls.

(2) *Owls and raptors that can be observed in the forest interior (name tracks 10-14).* As with the birds of the previous group, the austral owls inhabit the forests and are principally detected by their calls because they are most active at dusk and night. One of the diurnal birds of prey, *Accipiter bicolor chilensis,* is included in this section because it represents a unique case of a raptor that is specialized to hunt in the forest interior, silently and skillfully flying among the canopy.

(3) *Wetland birds, associated with riparian, coastal or bog habitats (name tracks 15-23).* When one wanders through the austral forests it is frequent to encounter courses of water, lakes, fjords, channels, rush thickets and other wetland habitats. In these habitats one can observe birds that live in riparian environments located inside or outside of forests. Hence, with the exception of the Dark-Bellied Cinclodes, none of these species are endemic to the South American temperate forest biome; they have wider geographical distributions within South America, or the Americas, and several of the species of this group of birds are migratory in their habits.

(4) *Birds of the forest margins, which are observed exclusively or principally along the border of forests and/or in adjacent, open habitats (name tracks 24-42).* These species are easier to detect visually given that they are found in open places. They are more generalist in their habitat use, and this group of birds includes both endemic and non-endemic species to the South American temperate forest biome. Among the latter, we find the only migratory bird that completely leaves the biome during winter: the White Crested Elaenia (*Elaenia albiceps*).

(5) *Raptors, which are usually seen perched on tall trees or soaring at great altitudes over the austral forests and a wide diversity of adjacent habitat types (name tracks 43-51).* They are mostly observed while soaring or flying, because they travel greater distances each day than do the previous groups of birds. Their geographic ranges tend to be extensive, as well. However, many of them depend on forest habitats to nest, perch and feed.

Bird names, and stories in the text and in the CDs. The description of each bird species in the text begins with its Yahgan, Mapudungun, Spanish, and English names. This order is repeated throughout the text and CD recordings. For Yahgan and Mapuche names, we only recorded those known or preferred by the Yahgan grandmothers Úrsula and Cristina Calderón, and by the Mapuche poet Lorenzo Aillapan, respectively.

The names are followed by a natural history and an ethnoecological narrative of the bird, which is complemented in 27 of the species by a recorded Yahgan or Mapuche story. The texts of the recorded stories on the CDs are indicated by the symbol $\odot$ The recorded Mapuche stories include 16 of the bird species, and are recorded on CD I, and the Yahgan stories are recorded in CD II. They include 10 species, plus one Mapuche and Yahgan story for the Austral Trush.

Regarding the transcription of the Mapuche stories into Mapudungun language, these were written in conjunction with Mr. Aillapan, who closely follows the Unified Mapuche Alphabet, proposed by linguist María Catrileo. The Yahgan terms recorded with the grandmothers Úrsula and Cristina Calderón were transcribed as closely as possible to an equivalent in Spanish pronunciation. Using the recordings of this guide, the Yahgan names are being transcribed by linguists to the International Phonetic Alphabet (IPA). The Yahgan and Mapuche names that are taken from bibliographic sources maintain the original format of the source.

In the final page for each species, we provide Yahgan, Mapuche, English and Spanish names that are documented from other ethnographic interviews or bibliographic sources. As you may have already noticed, some terms in the Introduction are written in color. This color coding identifies the language of the words and bird names:

> Blue is Yahgan;
> Red is Mapudungun;
> and green is used for the common names (including other species of flora and fauna) in English.

Distribution maps for each bird species indicate the geographic range of the bird, taking into account seasonal movements and/or degrees of abundance. These maps use the following color code:

- Pale green year-round resident
- Yellow summer range
- Light blue winter range
- Violet occasional visitor

Additionally, the general location of territories for the Yaghan and Mapuche communities are depicted on the maps.
Below each map, we offer concise ornithological information, including its sighting probability, and a table with a brief description of the bird's identification marks, preferred habitats, habits, diet, and its conservation status.

Sighting probability is based on a scale of increasing probability of observing the bird:
1 Rare or difficult to see
2 Seen occasionally
3 Seen frequently
4 Very abundant, seen in almost all field trips

Identification marks of each bird begins with its total length, followed by the diagnostics features of the species.

Conservation status of the birds considers the Endemic Bird Areas (EBA) defined by Birdlife International. The two main areas are EBA 060 (Central Chile), and EBA 061(Chilean Temperate Forests); for a few species we refer to EBA 059 (Juan Fernandez Archipelago). These three areas are listed by Birdlife International as priority areas for conservation due to their severe or major habitat loses. For species that have identified subspecies and for species with congeneric sister species in southern South America, we also analyzed the conservation status of the individual subspecies, or sister species.

Species with conservation concerns are indicated by a red circle, with a letter inside that notes the conservation risk category: CR = critically endangered, EN = endangered, V = vulnerable, NT = near threatened.

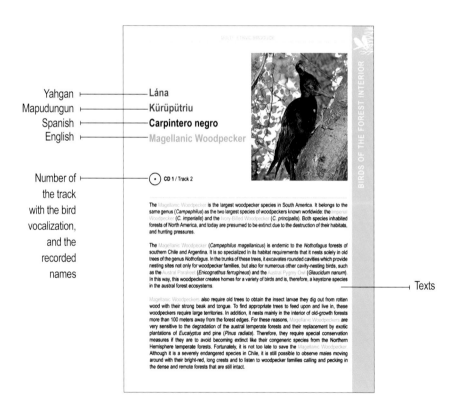

Yahgan ⊢——————— Lána
Mapudungun ⊢——————— Kürüpütriu
Spanish ⊢——————— **Carpintero negro**
English ⊢——————— Magellanic Woodpecker

Number of ⊢——————— ⊙ CD 1 / Track 2
the track
with the bird
vocalization,
and the
recorded
names

Texts

Yahgan or mapuche recorded story of bird (10 species have Yahgan stories, 16 birds have a Mapuche story, and 1 species has a Yahgan & Mapuche recorded story).

Story

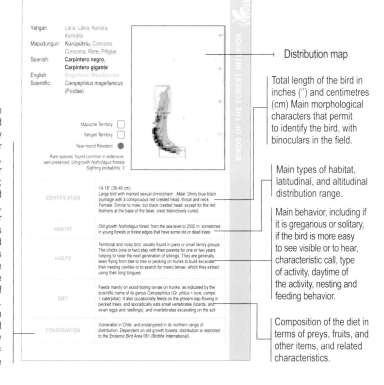

Distribution map

Total length of the bird in inches ('') and centimetres (cm) Main morphological characters that permit to identify the bird, with binoculars in the field.

Reference to endemism using the Endemic Bird Areas (EBA) defined by Birdlife International; for species with subspecies, the analysis is done for the individual subspecies; for species with related congeneric sister species, data are provided for both. General comments on conservation, and conservation status according to Birdlife International, IUCN, the Chilean Red Book of Fauna, and other sources. Species with conservation problems have a red circle with the category (En = endangered), V = vulnerable

Main types of habitat, latitudinal, and altitudinal distribution range.

Main behavior, including if it is gregarious or solitary, if the bird is more easy to see visible or to hear, characteristic call, type of activity, daytime of the activity, nesting and feeding behavior.

Composition of the diet in terms of preys, fruits, and other items, and related characteristics.

BIRDS
OF THE FOREST INTERIOR

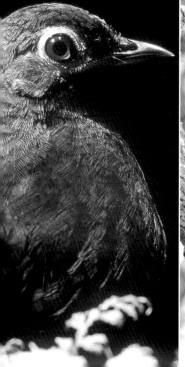

Lána
Kürüpütriu
Carpintero negro
Magellanic Woodpecker

(•) **CD 1** / Track 2

The Magellanic Woodpecker is the largest woodpecker species in South America. It belongs to the same genus (*Campephilus*) as the two largest species of woodpeckers known worldwide: the Imperial Woodpecker (*C. imperialis*) and the Ivory-Billed Woodpecker (*C. principalis*). Both species inhabited the forests of North America, and today are presumed to be extinct due to the destruction of their habitats, and hunting pressures.

The Magellanic Woodpecker (*Campephilus magellanicus*) is endemic to the *Nothofagus* forests of southern Chile and Argentina. It is so specialized in its habitat requirements that it nests solely in old trees of the genus *Nothofagus*. In the trunks of these trees, it excavates rounded cavities which provide nesting sites not only for woodpecker families, but also for numerous other cavity-nesting birds, such as the Austral Parakeet (*Enicognathus ferrugineus*) and the Austral Pygmy Owl (*Glaucidum nanum*). In this way, this woodpecker creates homes for a variety of birds and is, therefore, a keystone species in the austral forest ecosystems.

Magellanic Woodpeckers also require old trees to obtain the insect larvae they dig out from rotten wood with their strong beak and tongue. To find appropriate trees to feed upon and live in, these woodpeckers require large territories. In addition, it nests mainly in the interior of old-growth forests more than 100 meters away from the forest edges. For these reasons, Magellanic Woodpeckers are very sensitive to the degradation of the austral temperate forests and their replacement by exotic plantations of *Eucalyptus* and pine (*Pinus radiata*). Therefore, they require special conservation measures if they are to avoid becoming extinct like their congeneric species from the Northern Hemisphere temperate forests. Fortunately, it is not too late to save the Magellanic Woodpecker. Although it is a severely endangered species in Chile, it is still possible to observe males moving around with their bright-red, long crests and to listen to woodpecker families calling and pecking in the dense and remote forests that are still intact.

In territories that preserve large expanses of austral forests there will almost always be small families of woodpeckers marking their presence with the loud drumming noise of their pecking. On Chiloe Island this intense sound is associated with the chopping of the mythic *trauco's**hatchet. The Mapuche people of this island, the *Williche,* call this woodpecker *rere* or "rooster of the mount," because its wood-pecking percussion echoes throughout the mountains. The Mapuche people of the continental coasts of Temuco, the *Lafkenche,* highlight the color of this large woodpecker; hence, it is given the *Mapudungun* name of *kurüpütriu,* which describes it as a black (*kurü*) woodpecker (*pütriu*).

The large Magellanic Woodpecker or *rere* reaches the most southerly distribution of all the woodpeckers in the world, inhabiting even the forests of the Cape Horn Archipelago, where the Yahgan know it as *lana.* The name *lana* derives from the Yahgan word *lan* that means tongue, referring to the long and strong tongue that the woodpecker uses to extract the larvae. The Yahgan Grandmother Úrsula Calderón remembers that when she was a little girl she very much admired the skill with which the woodpeckers, or *lana,* pecked holes and extracted grubs or white larvae with their tongues, which they swallowed by throwing their heads back. Her father, Juan Calderón, told her the following story that taught her the kinship between birds and humans.

* *The* trauco *is a small mythic inhabitant of the forests of Chiloe Island. He has a human figure and dresses with branches of the vine* Luzuriaga radicans. *He carries a stone hatchet that he uses to gather fruits and vegetables, and glean food from rotten trunks. He walks through the forests pecking with the hatchet, like a* rere.

YAHGAN STORY **CD 2** / Track 10

The Magellanic Woodpecker or *lana* in the forests of the Cape Horn Archipelago accompanied the Yahgan women when collecting *dihueñes** in the forests. A Yahgan grandfather, Juan Calderón, said the origin of this beautiful bird of the austral forests goes back to ancestral times when birds were still humans.

In those remote times, a brother fell in love with his own sister and tried every kind of trickery to be with her and sleep together. His sister noticed this intention and eluded her brother each time to avoid taboo relations, but in the end she was of two minds: half of her wanted her brother and half of her did not.

The brother, who continued thinking of pretexts to bring her away from the *akar,* or hut, discovered one day giant, red *chaura* or Prickly Heath (*amai, Gaultheria mucronata*) berries in a forest gap and quickly went to tell his sister, "I have found enormous *chauras* in a place in the forest. You should go there and collect them." The sister took her basket and went into the forest, while her brother followed without anyone noticing and hid himself in a place where his sister had to pass. When she came sufficiently close, he embraced her, and they fell to the ground giving course to their love. When they stood up, they converted into birds and flew like *lana.* Since that time, they live together in the forests, and the brother carries over his head a red crest that reminds one of the color of those big *chaura* fruits.

**Edible endemic fungi from the genus* Cyttaria, *which grow only on the bark of* Nothofagus *trees.*

Yahgan: Lána, Lana, Kanára, Kankára

Mapudungun: **Kürüpütriu**, Concona, Concoma, Rere, Pitigüe

Spanish: **Carpintero negro, Carpintero gigante**

English: Magellanic Woodpecker

Scientific: *Campephilus magellanicus* (Picidae)

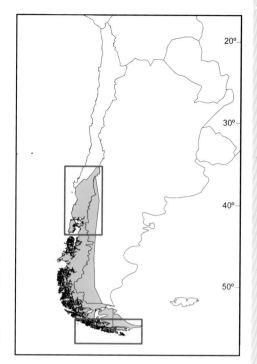

Mapuche Territory ☐

Yahgan Territory ☐

Year-round Resident ⬤

Rare species, found common in extensive, well-preserved, old-growth *Nothofagus* forests
Sighting probability: 2

IDENTIFICATION	14-18" (36-46 cm) Large bird with marked sexual dimorphism. *Male:* Shiny blue-black plumage with a conspicuous red-crested head, throat and neck. *Female:* Similar to male, but black-crested head, except for the red feathers at the base of the beak; crest distinctively curled.
HABITAT	Old-growth *Nothofagus* forest, from the sea level to 2,000 m; sometimes in young forests or forest edges that have some old or dead trees.
HABITS	Territorial and noisy bird, usually found in pairs or small family groups. The chicks (one or two) stay with their parents for one or two years helping to raise the next generation of siblings. They are generally seen flying from tree to tree or pecking on trunks to build and excavate their nesting cavities or to search for insect larvae, which they extract using their long tongues.
DIET	Feeds mainly on wood-boring larvae on trunks, as indicated by the scientific name of its genus *Campephilus* (Gr. *philus* = love; *campe* = caterpillar). It also occasionally feeds on the phloem sap flowing in pecked trees, and sporadically eats small vertebrates (lizards, and avian eggs and nestlings), and invertebrates excavating on the soil.
CONSERVATION	Vulnerable in Chile, and endangered in its northern range of distribution. Dependent on old-growth forests; distribution is restricted to the Endemic Bird Area 061 (Birdlife International). **V**

51

Túto
Tiftifken
Churrín
Magellanic Tapaculo

CD 1 / Track 3

From far away, the tiny Andean or Magellanic Tapaculo is heard when it emits its intense territorial call "*churin, churin, churin.*" The *Mapudungun* name *churin* repeats the call, and has been adopted by Spanish-speaking Chileans. As the *Lafkenche* poet Lorenzo Aillapan explains, "because of the regularity of the *churin's* vocalization it calls to mind a pocket-watch." Therefore, the Mapuche people of the coast or *Lafkenche* have metaphorically named this little bird the *triftrifken* or "pocket-watch bird." Even more, with its vocalizations the *triftrifken* marks the times of days and seasons, and in this way it helps maintain the rhythm of life in the temperate and sub-Antarctic forests, as expressed in the following verses by Aillapan.

The Magellanic Tapaculo - The Pocket-watch Bird	Tiftifkenkawün
Millenary bird that begins giving the hour from the daybreak, throughout the morning, to the noon-time and into the afternoon. It works by the light of day, and this it divides into four parts: the daybreak, the morning, the noon time and the afternoon, symbol of life, work, results and production, family, planting, harvest and animals Trutrif tif tif tif tif ken ken ken ken Trutrif tif tif tif tif ken ken ken ken!	Kuyfi üñüm rumel wülniekey rakiñ, mülen antü Epe wün, ella liwen, rangiantü, rupan antü amuley Re pelon mülem antü mütem küdaukey ta pu che meli kimfaluwi Wünlu, lüwengelu, rangiñ antü, ka tachi rupan antü Mongepiyum che, mülen küdau, rakiduam ka fill yallümün: Tachi pu mongeyel, ngangen, püram fün ketran, ka wera hulliñ. !Trutrif tif tif tif tif ken ken ken ken trutrif tif tif tif tif ken ken ken ken!

Nature, the cosmovision, the universe and relaxation,
great mountain range, plains, rivers, lakes, volcanoes and seas,
great spirit, the principal one, our four guardians,
that take care of the Mapuche world as an "Earthly Universe,"
that is the simple territory of the four winds,
that is the north, south, east and west without ends.
Trutrif tif tif tif tif ken ken ken ken
Trutrif tif tif tif tif ken ken ken ken!

I call myself Witranalwe, the treasure;
Anchümalleñ, female heat.
I am Dumpall, power of the sea; Püllüam, Mapuche spiritual force.

Four guardians of the Sacred Earth
tell the hour with the "pocket-watch bird" tiftifken…,
pocket-watch that indicates the human pair,
always in action and harmony with nature
Trutrif tif tif tif tif ken ken ken ken
Trutrif tif tif tif tif ken ken ken ken!

It is said that the native people are behind the times,
So I will set my watch forward.
Trutrif   tif tif tif tif trutrif   tif tif tif tif!

Neyen mapu, wenu püllü kimün, wallpmapu, peumatun ümag punlu
Fütra winkul lelfün, witrun leufü, leufü, degüñ, ka lafken fillem
Omfücha/omkude, wünen püllü, ka taiñ meli ñidolgen mapungelu
Küme kellulekelu witrañpüram pu che "nag mapu" mülewe mew
Feyta ñi fentren lof mapu tuwü külelu meli witran kürüf püle
Fey ñi felen Pikum, willi tripantü, konun antü püle, fill wallpan.
¡Trutrif   tif tif tif tif   ken ken ken ken
tru trif   tif tif tif tif   ken ken ken ken!

WITRANALWE pingey ülmen, ANCHIMALLEÑ pingey allangechi eñum
DUMPALL lafkenko ülmen, PÜLLÜAM pingey newen püllüngey

Rüf feyengün meli NGENGELU kellu witrañpüram chefengün
Alkütu penieyey rupalen antü tiftifken üñüm mew
Rakin triftrifken antüwe kimfaluwi mülen epu kurewen che
Rumel negüm kudau ka tuteuküley chew ñi Choyüngen:
¡Trutrif   tif tif tif tif   ken ken ken ken
trutrif   tif tif tif tif   ken ken ken ken!

¡Wall Mapuchengen piam pulay ka che kimü
Feula puy antü!
¡Trutrif   tif tif tif tif trutrif   tif tif tif tif!

The calls of the Magellanic Tapaculo are heard from far away, but it is difficult to see because, like the Chucao Tapaculo and the Black-Throated Huet-Huet, this small and dark bird lives immersed in the dense understory vegetation of the austral forests, preferably near streams. However, in contrast to its relatives, it sometimes ventures forth to the margins of the forest and is even found in dense brush habitat outside the forests, especially on the islands in the southern channels. Perhaps because of its ability to live in diverse habitats, the Magellanic Tapaculo is the only Rhynocriptidae to reach the world's southernmost forests. Although the Tapaculo family is tropical in origin, the Magellanic Tapaculo was first described based on a specimen collected in Tierra del Fuego in the 18th century; for this reason, *magellanicus* is its species name.

The Magellanic Tapaculo feeds on insects while it moves among the lower branches and the ground. In the understory of the forests, the Tapaculo also carefully builds its nest with fibers, roots, lichens, mosses, and small twigs in trunk cavities or between large branches. Its unique ability to build these nests contrasts with the other tapaculo species of the austral forests. Perhaps this skill of the Magellanic Tapaculo inspired the Yahgan story that honors this bird, *túto*, with its terrestrial lifestyle and poor flying capacity, for inventing the *anans*, or canoes. These canoes were made of *shapea* bark, or Magellan Evergreen Beech (*Nothofagus betuloides*), and allowed the Yahgans to navigate through the channels, bays and fjords of the Cape Horn Region.

YAHGAN STORY

 CD 2 / Track 8

An ancient Yahgan story tells that in the time of the ancestors, when birds were still humans, the Magellanic Tapaculo or *túto* was an old widower. He lived a long time with his son-in-law, who received the old man into his *akar* or hut. As the old man did not have other relatives, his son-in-law cared for him and maintained him during the last days of his life. However, the old man was always unsatisfied, even though the son-in-law took great pains, and every day went to the beach to collect him shellfish, meat and, occasionally, fruits from the forest. Nothing was to the old man's pleasure, and although he ate everything his son-in-law brought him, he always said, "I am already very old. I'd like to chew on something that I really like." So, the son-in-law looked for other things and brought something new each day, until one day he brought upon his return to the *akar* strips of *uri**, which the father-in-law chewed with great pleasure. He did not complain, but rather with great satisfaction said, "This is what I like."

When the son-in-law heard these words he sighed and said very happily, "At last, how happy and complacent my father-in-law is with the fibers of *uri* that I brought him. Now I will bring him many fibers. How could I have known that he would like them so much?" He went each day and always brought a bundle of *uri* that he put before his father-in-law. Tuto began to chew the fibers immediately, and after having carefully chewed a great quantity of fibers, he took pieces of bark and squeezed them with the fibers of *uri*, thereby inventing the *anans* or canoes that the Yahgan used to travel through the austral archipelagos. The old man made many canoes and gave them to the people. Since that time, one always needs *uri* to stick together the pieces of bark, so the canoes withstand the waves for a long time.

* Uri *is the Yahgan word for the fibers that can be extracted under the bark of the* Austral Beech (Nothofagus antarctica).

Yahgan:	**Túto,** Tútu, Toutou, Toutou-yakamouch
Mapudungun:	**Tiftifken**, Churin, Chercan, Chircan, Chochif
Spanish:	**Churrín,** Churrín andino, Para atrás
English:	Magellanic Tapaculo, Andean Tapaculo
Scientific:	*Scytalopus magellanicus* (Rhinocryptidae)

Mapuche Territory

Yahgan Territory

Year-round Resident

Common in native habitats with well preserved shrubby vegetation
Sighting probability: 1

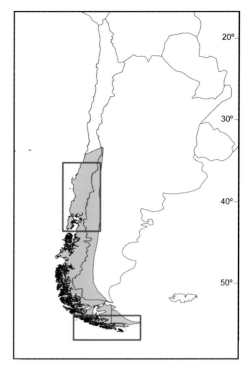

IDENTIFICATION	4-5" (10-13 cm) Small black to gray bird, and some individuals have an eye-catching white-silver spot on their forehead.
HABITAT	Old-growth forests with dense understory, and humid habitats with thick shrubby vegetation, especially along streams. From the sea level to the Andean Cordillera (3,000 m).
HABITS	Terrestrial, timid bird, moving among dense vegetation, close to the ground. Hence, it is difficult to see, but its loud calls reveal its presence.
DIET	Omnivorous diet, including invertebrates, and fleshy fruits, depending on seasonal availability.
CONSERVATION	The Magellanic Tapaculo is very sensitive to habitat degradation, and the introduction of exotic carnivorous mammals, such as cats, dogs, and the North American mink. Its distribution is restricted to the Endemic Bird Areas 060 and 061 (Birdlife International).

BIRDS OF THE FOREST INTERIOR

Wed-wed

Huet-huet

Black-Throated Huet-Huet

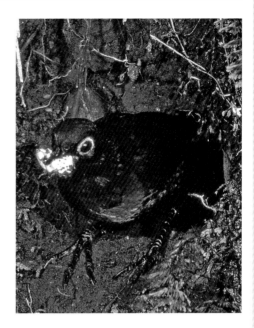

CD 1 / Track 4

The Black-Throated Huet-Huet is the largest Passerine bird that inhabits the temperate forests of austral South America. It lives in the interior of dense jungles, from the sub-Antarctic forests in the archipelago area of Wellington Island (49° S) to the Valdivian rainforests of the large Bío-Bío River (37° S). North of the Bío-Bío River until the Tinguiririca River (34.5° S), in the Mediterranean *Nothofagus* and sclerophyllous forests, lives its sister species, the Chestnut-Breasted Huet-Huet (*Pteroptochos castaneus*). Hidden among the understory, the Huet-Huet makes short flights and frequently is found scratching the soil in search of food that includes a great variety of invertebrates, seeds and fallen fruits. They also excavate large galleries in the earth that serve as nests and sleeping sites.

Dark colored, this Tapaculo species is difficult to see among the dense vegetation. Despite its ability to blend visually into its habitat, the presence of the Black-Throated Huet-Huet is signaled by its intense calls, especially its territorial call *"hoo-hoo-hoo,"* which is characterized by a series of descending notes that arise from the bottom of ravines or thickets of the southern rainforests, and are audible from great distances. Another call, its repeated warning *"wed-wed, wed-wed, wed-wed"* gives rise to its onomatopoeic *Mapudungun* name *wed-wed*, from which its Spanish names, *huet-huet* and *hued-hued,* are derived. This terrestrial bird also has distinct variations of its calls, which the Mapuche people of Chiloé Island, the *Williche*, interpret in order to forecast the weather. The call *"dehuet-dehuet,"* it will rain; the call *"mahoo-mahoo-mahoo- mahoo"* announces drizzle; when one hears the penetrating *"hoo, hoo, hoo..."* coming from the ravines, there will be sun.

In *Mapudungun*, *wed-wed* generally means "crazy" and, more specifically, a person with a joke flowering on his lips who knows that what he says is funny and dares to say, and show it. From far away in the ravines, the Black-Throated Huet-Huet sounds like the talking of a crazy person who accompanies travelers through the forests, telling curious stories that amuse them as they go.

57

BIRDS OF THE FOREST INTERIOR

Yahgan: out of bird's range

Mapudungun: **Wed-wed**,Hued-hued, Huahueta, Huid-huid

Spanish: **Huet-huet**, Hued-hued

English: Black-Throated Huet-Huet

Scientific: *Pteroptochos tarnii* (Rhinocryptidae)

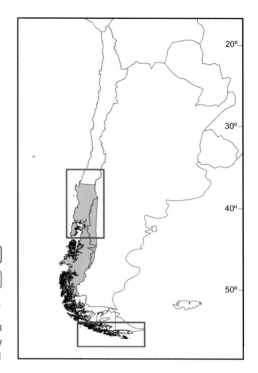

Mapuche Territory ☐

Yahgan Territory ☐

Year-round Resident ●

Relatively common in temperate forests of southern Chile and Argentina with well preserved understory
Sighting probability: 1

IDENTIFICATION	8-10" (22-25 cm). The largest Passerine bird in the temperate forests of South America. Its crown, rump, and belly are chestnut, while its back, throat, head and characteristic erect tail are blackish; its conspicuous eye-ring is whitish.
HABITAT	Native forests with well-developed understory, especially in dense patches of *Chusquea* (bamboo), and wet zones close to streams; from the sea level to the Andean Cordillera (1,500 m).
HABITS	Easier to hear than to see. Its loud vocalizations are heard from far away as a series of tones going down in pitch (hoo, hoo, hoo…), or as a rapid, onomatopoeic alarm call "wed-wed." It moves among the low branches of dense vegetation, and scratches the ground in search of food. It nests mostly on cavities of large trunks.
DIET	Omnivorous, terrestrial bird that feeds on fruits, seeds, insects, worms and other invertebrates, which it finds by scratching leaf litter with its strong feet.
CONSERVATION	Dependent on evergreen forests (37-49°S) with well preserved understory; distribution is restricted to the Endemic Bird Area 061 (Birdlife International).

Chukau
Chucao
Chucao Tapaculo

 CD 1 / Track 5

Repeating the sound of the *chukau's* calls, that peal forth from the understory of the forests, the *Mapudungun* name of this bird is onomatopoeic. However, the vocalizations of the *chukau* are varied, and the Mapuche people of Chiloe Island, or *Williche*, hear distinct messages in its different modulations, when they roam the forests. If the *chukau* sings to the right of the forest wayfarer, the expedition will be beneficial. On the other hand, if the call of the *chukau* is heard from the left, the fortune will be adverse. If the call of the *chukau* is constant and penetrating, then this announces rains and storms.

The Chucao Tapaculo, like the Black-Throated Huet-Huet, is an endemic species that characterizes the diversity of the temperate forests of southern South America. It also lives in forest interiors with dense canopy and a well-developed understory, where it is one of the most common birds. While moving among low branches or walking on the forest floor, the *chukau* forages for invertebrates and fruits fallen from trees, sometimes excavating the earth in its pursuit. This terrestrial bird collects twigs, herbs, and lichens to build its nests in the holes of trees or hillsides. Chucao males are very territorial, and stay in the same place for six years or more. They defend their territories by emitting loud calls, and by physically bumping chests with their competitors. Their short, loud, sharp and manifold vocalizations epitomize, aurally, the character of the forests of southern Chile and Argentina.

The Chucao Tapaculo is a beautiful bird with a bright orange throat and chest, yet it is difficult to see because it lives hidden among branches. The Mapuche people who inhabit the coastal forests of Temuco, or *Lafkenche,* tell that the *chukau* has the habit of walking secretly, bent over in the understory of the vegetation because it is like a beautiful maiden who was surprised by a young man when bathing under a waterfall or *trayenko* (*ko* = water, *trayen* = the naturally romantic sound of water).

MAPUCHE STORY

 CD 1 / Track 52

In *Lafkenche* territory, lovers court in the thicket of the forests, where many romances begin. The maidens go to the forest, also, to look for water or to take baths in the streams and waterfalls. It is told that the *chukau* bird has the habit of walking hidden in the thickness of the vegetation because long ago she was a beautiful maiden who was bathing under a waterfall or *trayenko*. She was surprised by a young man and felt embarrassed. Since that time this bird with its little tail standing up and its little legs towards the front wanders through the forests, like a maiden showing herself timidly, and at the same time coquettishly, between the foliage, where she emits her cries and call. In the old *Williche* weddings on Chiloé Island, which began with the abduction of the bride, the couple returned from the forest only once the *chukau* wished them all good luck with its call.

MAPUDUNGUN VERSION

Pu Lafkenche püle, üñawen peukey pu mawidantu mew, cheutañi müleken epuñpü e poyeuwün. Tachi pu malen komekeg rangintu mauwida meu müñetuaal mangiñko ka trayenko meu. Feyti weda chukau may ellkau miyaukey rangiñ karii lelfan ka fey witraley ñi pichi külen ka ñi epu namun rulalkuley, fey mag feyti allangechi we malen femngey müñetumekelu trayenko reke meu fey may perkey kiñe domokulmen wecheweutru. Fey may rumel müyaukey ñi am feyti malen lluru trekayawi rangiñ mawidantu pichike llükaley, welu duan niefuy chem, rangiñ tapül mawida, fey wefküley ñi mütrüm ül. Kayfimel pu kurewen williche CHÜLLEWE (Chiloé), feychi meu weñenge kefuy we malen, kurewen wüñotukefuy mawida meu feyti CHUKAU, Küme üL ülkantu wepum rumel.

Yahgan:	out of bird's range
Mapudungun:	**Chukau**, Chucao, Chucau, Tricau, Chiduco, Chukaw
Spanish:	**Chucao**
English:	Chucao Tapaculo
Scientific:	*Scelorchilus rubecula* (Rhinocryptidae)

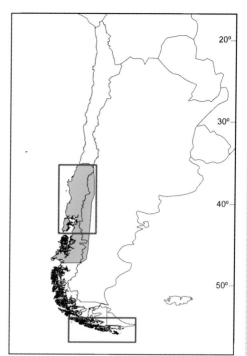

Mapuche Territory ☐
Yahgan Territory ☐
Year-round Resident ●

Common in Valdivian and Northern Patagonian
rainforests with well preserved understory
Sighting probability: 3

IDENTIFICATION	7-8" (18-19 cm) Gray head with a rufous eyeline, chest and neck. Its belly is gray with fine transverse black and white line, while its back and characteristic upright tail are brownish.
HABITAT	Native temperate forests with well developed understory, especially on dense bamboo (*Chusquea* spp.) patches, and humid areas near streams.
HABITS	It is easier to hear than to see. Its loud vocalizations, which emerge from the ground or low canopy, characterize the austral temperate forests. Due to its curiosity, if one stays still and silent the Chucao can be observed, and will approach visitors (it is more confident in the remote southern forests). It scratches the ground (as chickens do) in search of food, and it also carves its nests under dense vegetation.
DIET	It has a generalist diet, which includes fruits, seeds, insects, worms, and other invertebrates.
CONSERVATION	Species dependent on well-preserved native evergreen forests (37-49°S). Its distribution is restricted to Endemic Bird Area 061 (Birdlife International).

Tatajurj
Pelchokiñ
Comesebo grande
White-Throated Treerunner

 CD 1 / Track 4

The White-Throated Treerunner is endemic to the temperate forests of southern South America, and it is the sole representative of the genus *Pygarrhichas*. It lives in the forest interior, where it can be observed pecking trunks in the mid-level and high canopy like a little woodpecker.

This bird flits up and down the trunks and large branches of old trees, vigorously boring its own holes for nesting and feeding on the small insects and larvae that it finds. Its behavior shows us the importance of conserving the old and decaying trees that provide the basis of existence for the White-Throated Treerunner and many other living beings whose habitats are specific and scarce.

Although the White-Throated Treerunner is not large, it is conspicuous due to its prominent white throat and breast that inspired its scientific name *albogularis.* To hear this bird, you must pay attention to the short and low metallic sounds that resemble a rapid succession of drops. Outside the breeding season, they move around in forested areas, frequently forming mixed flocks with the Thorn-Tailed Rayadito (*Aphrastura spinicauda*).

This bird reaches the world's southernmost forested watershed on Horn Island (56ºS) in the Yahgan territory. Traditional Yahgan stories mention the White-Throated Treerunner, or *tatajurj*, accompanying the women or *kipa*, when they collected epiphytic fungi that grow on old trees of Evergreen Beech (*Nothofagus betuloides*), High Deciduous (*N. pumilio*) and Low Deciduos (*N. antarctica*) Beech. Like the White-Throated Treerunner, these fungi of the genus *Cyttaria, katran* in Yahgan language, are endemic to *Nothofagus* forests. The Yahgan Grandmother Cristina Calderón still remembers how much she enjoyed watching the *tatajurj* creeping up and down along the tree trunks.

YAHGAN STORY

CD 2 / Track 9

When I was a girl in Robalo, I always liked to go into the forest looking for twigs, firewood for grandfather Wisch* and Grandmother Hamumo, and in the forest I always saw the *tatajurj*. I liked them a lot. Sometimes I sat down for a long while to watch them, how they went up and down again and again, pecking the trunk. There were always two of them going together, the couple of *tatajurj*. I liked them so much that I was late in bringing the grandparents the firewood, and when I came back from the walk into the forest, they asked me, "Why are you so late?" And I told them that I did not find any firewood. Therefore, I had come late. I didn't tell them that I was watching these little birds.

Grandfather Wisch (or Whaits) told several traditional Yahgan stories to the anthropologist Martin Gusinde, who wrote the largest Yahgan ethnographic record. According to the Yahgan custom, the children took care of elderly people, bringing them water and firewood. Today, Robalo Bay is geographically a part of the Omora Ethnobotanical Park, a nature reserve where biocultural research, education, and conservation programs are conducted. These programs include the preparation of educational materials such as this book.

Yahgan: **Tatajurj**, Tatajúrj

Mapudungun: **Pelchokiñ**, Pishonquillu

Spanish: **Comesebo grande**, Comesebo, Picolezna, Carpintero carmelito

English: White-Throated Treerunner

Scientific: *Pygarrhichas albogularis* (Funariidae)

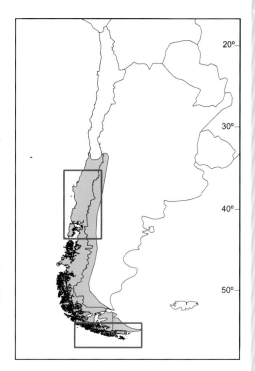

Mapuche Territory ⬚

Yahgan Territory ⬚

Year-round Resident ⬤

Common in well preserved old-growth forests
Sighting probability: 2

IDENTIFICATION	6-8" (14.5–17 cm) Conspicuous white throat and chest, which contrast with dark brown back, neck, and head. Rufous lower back and tail. Long, dark beak, slightly curved upwards.
HABITAT	Diverse types of old-growth forests, from the sea level to the Andean Cordillera (2,000 m).
HABITS	Agile creeper on trunks and large branches. Restless bird, sometimes moves with its head downwards, but generally ascending spirally around trunks, removing pieces of bark and emitting characteristic short, high-pitched sounds.
DIET	It feeds on insect larvae and adults that it captures over or below the bark, and very occasionally on the ground, especially on remote islands (such as the Cape Horn archipelago) that lack terrestrial predators.
CONSERVATION	*Pygarrhichas* is a monotypic and endemic genus; its distribution is restricted to Endemic Bird Area 061 (Birdlife International).

65

Tachikáchina
Pidpidwiriñ
Rayadito
Thorn-Tailed Rayadito

⊙ **CD 1** / Track 7

BIRDS OF THE FOREST INTERIOR

Aphrastura is another genus endemic to the forests of Chile and Argentina. This genus possesses two species: the Masafuera Island Rayadito (*A. masafuera*) which lives in the forests of the Juan Fernández Archipelago and is very endangered due to habitat destruction and introduced exotic predators; and the Thorn-Tailed Rayadito (*A. spinicauda*) which is found along the whole range of the *Nothofagus* forests, from Cape Horn to the relict coastal forests of Fray Jorge (31°S).

The Thorn-Tailed Rayadito is a little bird with white throat and breast, and yellowish-brown back plumage. Its common Spanish name *rayadito* (little striped one) is derived from the black lines that cross its crown, forehead, and wings. Its English (Thorn-Tailed) and scientific (*spinicauda*) names derive from its prominent tail, which has protruding outer feathers with terminal points.

Thorn-Tailed Rayaditos move tirelessly among branches of the canopy and understory and on trunks, where they peck the bark, looking for larval and adult insects. They nest in holes of live or dead trees and stumps, and are able to pass through very narrow openings to enter their nests. The nest provides protection from predators and contributes to their success in *Nothofagus* forests, where they are one of the most abundant bird species.

Thorn-Tailed Rayaditos are one of the most easily detected and seen birds in the interior of austral forests. They are very curious, and when you enter the forests they come close and emit constant noisy vocalizations that give them their onomatopoeic Yahgan name *tachikáchina*. The Yahgans appreciate *tachikáchina* because they announce the presence of any other animal or person. As Grandmother Úrsula Calderón tells us, the Thorn-Tailed Rayadito is a watchguard that detects and announces dangers.

YAHGAN STORY

 CD 2 / Track 7

Tachikáchina, or the Thorn-Tailed Rayadito, is a bird that sings during the day in the mountains to warn that somebody is hiding, a bad man, a witch-doctor. He tells the one who comes walking about the presence of others, also about any dog, cat.... well, about any hidden thing. Their cries make you afraid when they sing together, *tsch, tsch, tsch*, as they alert; nothing good.

Yahgan: **Tachikáchina,** Tečikášina, Tulératécix

Mapudungun: **Pidpidwiriñ,** Pidpid, Pishpish,Yiquiyiqui

Spanish: **Rayadito,** Raspatortillas, Yiquiyiqui

English: Thorn-Tailed Rayadito

Scientific: *Aphrastura spinicauda* (Funariidae)

Mapuche Territory

Yahgan Territory ☐

Year-round Resident ⬤

Very common in well preserved forests, throughout the geographic range of South American temperate forests

Sighting probability: 4

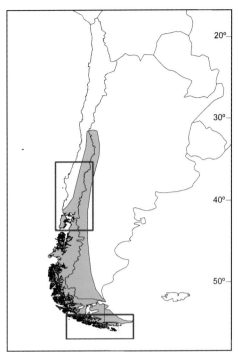

IDENTIFICATION	5-6" (13–15 cm) Black crown and sides of head, separated by a conspicuous yellowish eyeline from the base of the beak to the nape. Whitish throat, chest and belly, rufous back, and black wings with two yellowish bars. The long tail is very distinctive, with rufous central feathers and black lateral feathers whose rachises are prolonged naked (without barbs at their tips) resembling thorns.
HABITAT	Diverse types of forests, including relict forests in Fray Jorge National Park (30°30'S), Central Chilean sclerophyllous woodlands, monkey-puzzle tree, Valdivian, Northern Patagonian, and sub-Antarctic forests, and also sub-Antarctic shrublands and tussock grasslands in Cape Horn and Diego Ramírez islands (56°30'S).
HABITS	Tireless and noisy bird, creeps acrobatically on trunks and branches. Usually moves in small flocks, sometimes associated with other species forming mixed flocks, especially with *Pygarrhichas albogularis* during autumn and winter. Quite tame and curious bird, when visitors hike along trails the Rayaditos come close to them emitting characteristic vocalizations from the canopy. It builds nests in natural, narrow cavities in trunks, and occasionally in cliffs or rural houses. During the reproductive season, they can be observed continuously bringing insects to the nests hidden inside the trunks.
DIET	It feeds on coleopteran, and other insect larvae and adults; it occasionally eats fruits and seeds.
CONSERVATION	Endemic genus of south-western South America, with two species: *A. spinicauda* restricted to the Endemic Bird Areas 061 (Birdlife International), and the critically endangered *A. masafuera* restricted to Alejandro Selkirk Island in the Juan Fernandez Archipelago (Endemic Bird Area 059).

Kinan
Choroi
Cachaña
Austral Parakeet

 CD 1 / Track 8

The parrot family has a near-worldwide distribution, but the majority of the species are concentrated in tropical latitudes. For that reason, when the French naturalist Bougainville referred to parrots that lived in Tierra del Fuego, scientists of the 18th century did not believe him. Later, the expedition of *HMS Beagle* and the naturalist Charles Darwin confirmed that the Austral Parakeet is indeed an exceptional parrot, which reaches the southernmost forests of the planet, inhabiting even the islands of Cape Horn. Moreover, the Austral Parakeet remains there year-round. Lucas, son of the Anglican missionary Thomas Bridges wrote, "these birds do not migrate, and in winter, although it might seem misplaced, the parrots can be seen in the canopy of trees covered by snow." Bridges also wrote that the Fuegian Indians, Yahgan and Selknam (or Ona), did not like the Austral Parakeet because they felt deceived by these noisy birds, which gave the guanaco hunters away to their prey as they followed them through the forest. For this reason, the Austral Parakeet, or *kinan* in Yahgan language, was believed to be a spy, working for the guanacos. This belief by the Fuegian Indians does not surprise scientists, because a flock of austral parakeets usually has a sentinel bird that alerts its companions to the presence of predators or other danger with loud alarm cries.

Covering the entire range of *Nothofagus* forests in southern Chile and Argentina, gaudy flocks of Austral Parakeets fly over, or perch in, the canopies of Evergreen Beech (*Nothofagus betuloides*), Tall Deciduous Beech (*N. pumilio*) and other trees. They eat seeds voraciously, and nest in holes inside old tree trunks. Between the leaves and branches of *Nothofagus* trees, one can observe their showy bright green-yellowish plumage with rust colored tails, which is where its scientific name, *ferrugineus*, comes from. Austral Parakeets fly very skillfully through the canopy. They avoid obstacles and hunters. Denoting this ability, the *Mapudungun* name *kachaña* means "to elude." and in Chile people use the term "*cachaña*" (the Chilean name for the Austral Parakeet) to describe a person who eludes others with rapid and elegant movements.

Austral Parakeets are also admired for their capacity to imitate and speak. On Navarino Island, the Yahgan Grandmother Cristina Calderón remembers that when she was a child and lived in Robalo Bay, there was a *kinan* that called her uncle Juan by his name. Later, another *kinan* lived with Grandmother Candelaria in Villa Ukika, where the Yahgan community lives today. That bird had the ability to imitate the cackling of the hens. In Antilhue, near the city of Valdivia, people feed captive Austral Parakeets with bread soaked in wine to teach them bad words. In country houses, it is common to find captive parakeets. If they are treated kindly, Austral Parakeets like the families who befriend them--but demonstrate a great antipathy if they are treated poorly. On Navarino Island, however, Grandmother Cristina prefers not to have captive *kinan* because they suffer in cages and prefer to fly without hindrance. Birds enjoy freedom, she says, as much as we human beings do.

In the Chilean forests further to the north lives another parrot species of the same genus: the Slender-Billed Parakeet (*Enicognathus leptorhynchus*). It is much less common than the Austral Parakeet, and might well be called the "Chilean Parakeet" since it is found exclusively west of the Andes Cordillera. The Slender-Bill is similar to the Austral Parakeet, but can be distinguished by its much longer bill, larger body size (40 cm), and its red forehead and lore. Its range of distribution is also much more restricted. The Slender-Billed Parakeet lives mainly between the Bío-Bío River (38ºS) and Chiloé Island (42ºS), although it can be found occasionally between Valparaíso and Aysén. In the mountains of Temuco in the Allipen Valley is the town of Choroico; its name means water (*ko*) of the *choroi*. On Chiloé Island the Choroihue River means place (*we* or *hue*) of the *choroi*. This ornithological toponymy expresses how abundant this parrot species was in the Mapuche territory. However, their populations have declined due to habitat destruction, hunting and because, like the Chilean Pigeon, it was affected by a fatal disease produced by the Newcastle virus in the middle of the 20th century.

The *choroi* especially typifies the Monkey-puzzle Tree (*Araucaria araucana*) or *pewen* forests, where it feeds on the cones that it opens without difficulty, using its sharp, hooked bill. The Mapuche people of these forests, the *Pewenche*, also eats the seeds contained in the cones of the *pewen*. *Pewenche* elders inhabiting the Andean zones of the Galletue and Icalma lakes (where the Bío-Bío River is born), still remember when, before they were hunted with rifles, *choroi* were so abundant they clouded the sky. In March of each year, hundreds of birds landed on *pewen* trees to eat almost all their fruits.

Near the ocean, where the *Lafkenche* live, the poet Lorenzo Aillapan also remembers that flocks of *choroi* were very abundant. They were so numerous and voracious that they caused farmers to erect scarecrows, because when flocks of this parrot landed on corn or wheat farms, they left nothing for the harvests. Consequently, *choroi* were intensively persecuted and hunted, contributing to their severe population decline. Today, both the Monkey-puzzle Tree and the Slender-Billed Parakeet are endangered, endemic species of southern South America. We hope that acquitance with the history of the *pewen* and the *choroi* can stimulate people to appreciate and preserve these ancient trees and birds, as well as the ecosystems and cultures with which they coexist.

Yahgan: **Kinan,** Kīnan, Kínan

Mapudungun: **Choroi,** Kachaña, Rawikma, Cachuña, Tricau, Rawilma, Chilkeñ

Spanish: **Cachaña,** Catita austral

English: Austral Parakeet, Chilean Green Parrot

Scientific: *Enicognathus ferrugineus* (Psittacidae)

Mapuche Territory

Yahgan Territory

Year-round Resident

Occasional Visitor

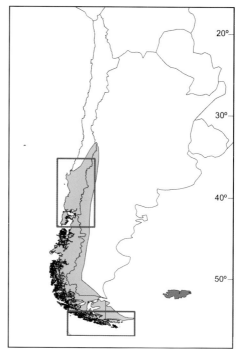

Common in *Nothofagus* forests, from the sea level to the Andean Cordillera
Sighting probability: 3

IDENTIFICATION	12-14" (31-36 cm) Parrot with a small gray bill, and a green plumage (darker on upperparts) with some redish feathers on the forehead, lorum and belly, and a distinctictive long, pointed tail ferrugineous in color.
HABITAT	Native forests, especially those dominated by *Nothofagus*, from their northern border at 34°S to their southern end in the Cape Horn archipelago (56°S).
HABITS	Gregarious, flying over or perching on the canopy in noisy flocks, which occasionally feed on the ground. They nest on large trunks, in natural cavities or those built and later abandoned by pairs of the Magellanic Woodpecker (*Campephilus magellanicus*).
DIET	To subsist in the southernmost forests of the world, the Austral Parakeet exhibits a broad vegetal diet, which varies with the seasons of the year. In early Spring, this parrot feeds largely on the canopy of *Nothofagus* trees which have fresh leaf buds, and great quantities of wind-pollinated flowers whose stamens are loaded with protein rich pollen. In late Spring, the flocks of parrots feed on firebush (*Embothrium coccineum*) flowers that are rich in nectar and pollen. In summer they eat lipid-rich *Nothofagus* seeds and berries of other species of trees. During winter their diets also include buds and leaves of mistletoe species (*Misodendrum spp.*), as well as epiphytic fungi of the genus *Cyttaria* that grow on trunks and branches of *Nothofagus* trees.
CONSERVATION	Genus endemic to south-western South America, with only two species: *Enicognathus ferrugineus* restricted to the temperate and sub-Antarctic forests of Chile and Argentina in the Endemic Bird Area (EBA) 061 (Birdlife International), and *E. lepthorhynchus* restricted to the Chilean temperate forests, mostly between 37 and 43°S (EBA 061).

73

BIRDS OF THE FOREST INTERIOR

Epukudén
Colilarga
Desmur's Wiretail

(•) **CD 1** / Track 9

Desmur's Wiretail is the only species of the genus *Sylviorthorhyncus*, and it is endemic to the southern temperate forests of Chile and Argentina. Thanks to its long tail, which can measure twice the bird's body length, it is an eye-catching species. There is no more of an amazing experience than to see it fly between the branches with its swaying tail, especially if it flies from one thicket of *küla* or bamboo (*Chusquea sp.*) to another. Its *Mapudungun* name *futrakülen* refers to the bird's long *(futra)* tail feathers *(külen)*.

On Chiloé Island, where the *Williche* people live, the tradition exists of keeping a *futrakülen* tail feather in books to better memorize and understand their contents. For this reason, the *futrakülen* or "long-tail" is known in the Chiloé Archipelago as "the remembers-all" (*el memorioso*), and students not only carry the tail feathers in their textbooks, but they even sleep with these books underneath their pillows to memorize and learn their lessons better. The other *Mapudungun* name *epukuden* might be related to this belief, since it means to foretell twice (*kuden* = foretell, *epu* = two).

For the students, as well as for birdwatchers the Desmur's Wiretail are difficult to find, hence their feathers are a treasure if one is found. This bird lives hidden in the thickness of the understory and its colors are not very eye-catching, being yellowish-brown on the under parts and reddish-brown on top. This bird hunts and feeds on insects among the branches of the *küla* or bamboo thickets. Its presence is detected more frequently by its call, formed by reiterated *ye-ye-ye-ye*, or by its prolonged territorial trill that emerges from the thickets, similar to the song of the Tufted Tit-Tyrant (*Anairetes parulus*). Among these thickets, their presence is also reveled by their nests composed of *küla* roots or leaves, which are easily recognized by their characteristic spherical shape with a single lateral entrance.

Yahgan: out of bird's range

Mapudungun: **Futrakülen, Epukudén,**
Changkülen, Changkihue

Spanish: **Colilarga**, Cola de paja,
Memorioso

English: Desmur's Wiretail,
Rusty Fern Tail

Scientific: *Sylviorthorhynchus desmursii*
(Furnariidae)

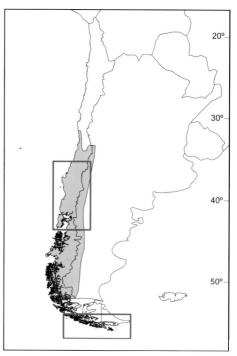

Mapuche Territory ☐

Yahgan Territory   ☐

Year-round Resident ●

Rare, found almost exclusively in forests with well
developed understory and patches of bamboo
Sighting probability: 1

IDENTIFICATION	8.5-9.5" (22-24 cm) Little bird with an unmistakable tail, having two very long central feathers (2/3 of the total length of the bird) and two pairs of external feathers, shorter in length. Rufous brownish plumage on the upper parts and yellowish brown on the lower parts. Head with reddish frontal forehead and crown, and a distinct whitish eye brow.
HABITAT	Native forests with well developed understory, especially in forest gaps and edges with dense patches of bamboo (*Chusquea* spp.), and wet shrubby areas close to streams. Found from the relict coastal fog-dependent forests in Cerro Santa Inés (32°S) to the sub-Antarctic rainforests south of Wellington Island in Magallanes (51°S), from the sea level to the Andean pre-cordillera (1,200 m).
HABITS	Easier to hear than to see because it spends most of its time moving among the low branches of dense vegetation. However, when it feels threatened it makes a repeated binary high-pitched call and/or long ascending trill that reveals its presence. It builds a ball shaped nest made of mosses, grasses, and plant fibers (especially bamboo leaves) lined with soft feathers, with an entrance on the side, and well hidden in the dense vegetation close to the ground.
DIET	Feeds mainly on insects, which it catches during flight; it also eats seeds (especially graminea).
CONSERVATION	Monotypic, endemic genus restricted to Endemic Bird Area 061 (Birdlife International).

OWLS AND FOREST
INTERIOR BIRDS OF PREY

Kuhúrj
Kong kong
Concón
Rufous-Legged Owl

 CD 1 / Track 10

The Rufous-Legged Owl is a close relative of the Northern Spotted Owl *(Strix occidentalis)*, who helped to motivate the conservation of the old-growth forests of the Pacific Northwest of the United States and Canada. In the Northern Hemisphere or the Southern Hemisphere, these owls prefer to inhabit old-growth temperate forests with dense canopy cover. From the complex vegetation structure of the austral primary forests, sometimes incorrectly called "overly mature forests," emerges the powerful, prolonged call *kong kong kong…* which is usually heard around midnight and is the inspiration of the onomatopoeic *Mapudungun-Lafkenche* name: *kong-kong.*

The Rufous-Legged Owl is identified by its varied nocturnal calls and by its brownish face with large concentric rings around its eyes. From within the dense foliage it usually emits a short call *kuhúrj* or a short, but strong, echoing shout *có-có-có*. The first sound led to its onomatopoeic Yahgan name *kuhúrj*; the second to its *Mapudungun-Williche* name *coa* or *có*. The *Williche*, who are the Mapuche people that live on Chiloé Island, assign distinct meanings to this owl's calls, depending on the form. If the *coa* only shouts once, it calls for good harvests and fortune, but if it calls many times, it announces bad luck. The "remedy" to change this message of bad luck consists of hanging salt over the wood stove.

For the *Lafkenche*, inhabiting the forests and farmlands along the continental coastal areas of the Mapuche territory, the *kong kong* is a beloved bird. When a thief, or *weñefe*, comes near with intentions of stealing a domesticated animal, the *kong kong* forewarns with calls that imitate a cackle, moo, neigh or bleat. As expressed by the *Lafkenche* poet Lorenzo Aillapan, this owl serves as an "alarm" for the communities.

OWLS AND FOREST INTERIOR BIRDS OF PREY

MAPUCHE STORY

 CD 1 / Track 53

For the Lafkenche, the *kong kong* is a nocturnal bird, very beloved for its ability to imitate other birds and animals, like chickens, cattle, horses or sheep. When some robber or *weñefe* comes close with the intentions to steal some of these domestic animals, the *kong kong* shouts. If it is going to steal a cow, it shouts ....*nruuu*. If it is going to steal a horse, it shouts ....*ihhhiiiii*. If it is going to steal a sheep, it shouts ....*meheeeee*

MAPUDUNGUN VERSION

Tachi pu LAFKENCHE, rume ayipoyekefi tachi KONGKONG punüñüm feyta inarume pikantufilu puwera üñüm ka fentren kulliñ, femngechi achawall rume, waka rume, kawellu rume ka ufida rume. Feyti mülekelu WEÑEFE fey witramealu fill wera kulliñ ñomngelu. Fey pipingey wakeññün KONGKONG.
Weñengeyal WAKA..... m u u u u – m u u u u.
Weñengeyal kawell i i i i . ji- ji – ji –ji –ji
Weñengeyal ufida   me – e – e – e – e

Yahgan: Kuhúrj

Mapuche: **Kong kong,** Concon, Coa, Ñeque, Colcón

Spanish: **Concón,** Lechuza bataraz

English: Rufous-Legged Owl

Scientific: *Strix rufipes*
(Strigidae)

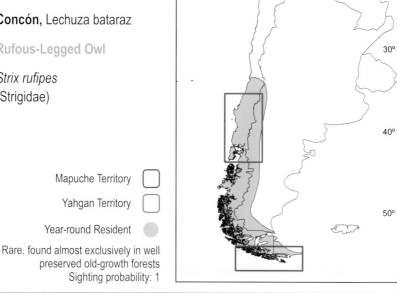

Mapuche Territory ☐

Yahgan Territory ☐

Year-round Resident ●

Rare, found almost exclusively in well
preserved old-growth forests
Sighting probability: 1

IDENTIFICATION	13-15" (33-38 cm) Robust, large headed owl with no ear tufts. Facial disk with black border, concentric bands of dark plumage, white eyebrows, lorum and mustaches, and dark eyes. Upper-parts blackish-brown finely barred with white, and under parts whitish densely barred with black. Characteristics rufous thighs.
HABITAT	Associated with old-growth forests from the sea level to the Andean Cordillera (2,000 m), from sclerophylous forests in Central Chile (33°S) to the sub-Antarctic rainforests on Hoste and Navarino islands (55°S) south of Tierra del Fuego.
HABITS	Easier to hear than to see. Nocturnal and cryptic owl, which presents diurnal activity in the cold old-growth forests of Navarino Island in the Cape Horn Biosphere Reserve, where they are cavity nesters that frequently use holes built by the Magellanic Woodpecker (*Campephilus magellanicus*) in large trees of the genus *Nothofagus*.
DIET	Mostly rodents and other small mammals (especially arboreal rodents and marsupials), but includes also birds, reptiles, amphibia, insects, and other invertebrates in its diet.
CONSERVATION	Recent analyses identify *Strix rufipes* as the Chilean raptor species with the highest conservation priority. Endemic to the South American tempetrate forests biome, Endemic Bird Area 061 (Birdlife International), a priority area for bird conservation at the world level.

OWLS AND FOREST INTERIOR BIRDS OF PREY

OWLS AND FOREST INTERIOR BIRDS OF PREY

Yahutéla
Tukuu
Tucúquere
Austral Great Horned Owl

(•) **CD 1** / Track 11

Distinguished by the horn-like tufts on its head that gives it its English name, the Austral Great Horned Owl is the largest of the austral owls. Its general coloration is brownish-gray with flecks of yellow, and black horizontal bands on the underside. In the austral landscapes, it can be seen perched on branches of tall trees or flying in search of rodents, rabbits and occasionally a bird or insect. It emits sonorous, merry whistles when it disputes its territory or mate. In contrast, when it is perched on a tree the Austral Great Horned Owl emitis loudly its characteristic call *tukuuhuhu, tukuuhuhu...* This the deep voiced ululation from which it gets its *Mapudungun* onomatopoeic name *tukuu*. When the voice of the *tukuu* is heard in the night, it seems to invoke a dense fog or *trukur* that makes the wayfarer become lost.

The Austral Great Horned Owl lives in diverse forest types and open environments, and even reaches the Cape Horn archipelago, where it receives the Yahgan name *yahutéla*. In the sub-Antarctic forests of Navarino Island, south of Tierra del Fuego, the Austral Great Horned owls are abundant, and according to an ancient Yahgan story, these birds were hunters who turned themselves into owls.

In ancestral times when birds were still humans, a Yahgan child lost his father and went to live with his mother and uncle. The uncle and the other men of the camp were very selfish, and never gave good food to this boy, and he became very skinny. One day, when she realized her son was so tired and hungry, his mother made him a pair of leather sandals, or *kili* in Yahgan language, and told him to go off by himself and hunt guanacos *(Lama guanicoe),* or *amara*. After crossing a great mountain, he killed many *amara* and brought a large one back to the camp site. His mother was very happy and began to eat huge chunks of good meat. Except for her, no one believed that the skinny little boy was capable of killing and transporting the *amara* to his hut, or *akar*. The next day the incredulous men, or *yamana*, accompanied the boy to the place where he left the other dead *amara*, on the other side of the high mountain where none of the *yamana* had been able to reach before. All were surprised to see the hunted animals and each manshouldered one to carry back to the camp. However, crossing the great mountain, the men became very tired because the *guanacos* they carried became heavier and heavier with every step. Only the skinny little boy was able to walk rapidly with his load and to return early to his *akar* where his mother was waiting. In turn, the egotistical men advanced very slowly. Under the weight of their burden they were transformed into owls, and arrived to the camp very late at night without their *amaras*. They could not enter into their *akar* where there relatives live. From that time to to our own, the *yahutéla* live in the forests, or *ashuna*, and come around the *akars* of the Yahgans at night, calling *tukuu, tukuu, tukuu* …

Yahgan:	**Yahutéla,** Kuhúrux, Yohutela, Yapoutéla
Mapudungun:	**Tukuu,** Nuco, Tuco, Ñacurutú, Raiquen
Spanish:	**Tucúquere**, Búho, Ñacurutú
English:	Austral Great Horned Owl
Scientific:	*Bubo magellanicus* (Strigidae)

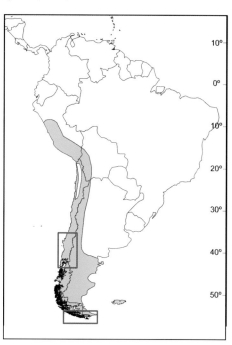

Mapuche Territory ▢

Yahgan Territory ▢

Year-round Resident ⬤

Inhabits a variety of forest and other habitat types
Sighting probability: 2

IDENTIFICATION	17-19" (43-49 cm) Large horned owl, with conspicuous yellow eyes and long auricular tufts that resemble horns. Throat and collar are white, separated by a brown semicollar. Pale brown facial disc with a black border. Upper parts are grayish brown, and underparts are pale gray-brown with narrow dark bars.
HABITAT	More common in open forests, woodlands, and park areas, but also found in shrublands, high Andean puna, and farmlands, from central Peru (12°S) to Navarino Island (55°S), and other islands of the Cape Horn Biosphere Reserve, south of Tierra del Fuego.
HABITS	Nocturnal owl that frequently perchs exposed, and presents some diurnal activity in the cold sub-Antarctic forests of Navarino Island. Nesting pairs are very aggressive defending their nest, and it could be dangerous to approach them.
DIET	Mostly small mammals, including rabbits and hares, but it also feeds on reptiles, invertebrates, as well as large and small birds; indeed, flocks of small birds vocalize nervously when the large owl is around.
CONSERVATION	Tens of individuals are kept in rehabilitation centers, due to hunting pressures by farmers who do not understand the ecological services that these owls provide by controlling populations of rodents and other organsims.

OWLS AND FOREST INTERIOR BIRDS OF PREY

OWLS AND FOREST INTERIOR BIRDS OF PREY

Sírra
Chiwüd
Lechuza blanca
Barn Owl

(·) **CD 1** / Track 12

In the countrysides, the Barn Owl is known as "monkey face" or "cat face," because its face possesses a type of white disk in the form of a heart that contrasts with the yellowish-brown color of the rest of the head. Its flight is absolutely silent. However, upon flying it frequently emits a raspy, strident call from which it receives its *Mapudungun* name: *Chiwüd*. As in the case of the other owls, the name that the *Mapuche* people give to the Barn Owl is onomatopoeic. The word *chiwüd* imitates the characteristic squeal or hiss of the Barn Owl.

The beliefs and practices associated with *chiwüd* vary markedly across the *Mapuche* territory. For the *Lafkenche* people inhabiting the coastal forests and farmlands, this owl is an "extravagant bird" because it bandies about, giving bad directions with its squeal to wanderers who try to make their way through dense fog. In the *Lafkenche* territory, it is told that, when travelers come upon a *chiwüd* in the road, they are sure to become lost. Only by putting their clothes on backwards, including their right shoe on their left foot and their left shoe on their right foot, can those deceived by the *chiwüd* expect to return to their proper path.

For the *Williche* inhabiting the broadleaf evergreen rainforests on the Chiloe archipelago, the call of the *chiwüd* might bring bad weather, and to counteract this, people throw ashes into the air. In the ancient *Pikunche* territory in the sclerophyllous forests of Central Chile, it is said: *chiwüd weda femkey anüpayüm*, "when the Barn Owl perches on a house it brings a bad omen." On the other hand, the Mapuche and other farmers value the Barn Owl because they frequently see the owl hunting mice at dusk. They applaud its excellent sense of hearing, that even enables it to hunt mice under snow; consequently, *chiwüd* is beneficial to agriculture.

Barn owls are largely sedentary. If a pair feels safe and has food, then they do not change sites during their whole life. For their nests they use tree cavities or the attics of sheds, houses, churches or other buildings. They live and nest in a variety of habitats, such as evergreen and deciduous forests, fields, rural areas and even cities.

OWLS AND FOREST INTERIOR BIRDS OF PREY

These owls are well known throughout most of the world, and in the Americas they live from Canada to Navarino Island in the Cape Horn region inhabited by the Yahgans. For the Yahgans the Barn Owl or *sirra* is associated with water. *Sirra* is the wise grandmother of the hummingbird *omora*, and participated in the creation of streams and rivers of Navarino Island. Today, still the Yahgans associate the sound that *sirra* makes when it flies at night with the sound of the streams or rivulets of water. As the Grandmother Úrsula Calderón tells:

YAHGAN STORY

CD 2/ Track 6

"What bird is this," I asked my grandfather. "It is the Barn Owl called *sirra*; there are many of them at Santa Rosa and Mejillones'. You hear them at night *shhh, shhh, shhh*." Perhaps they sleep during the day. I saw a *sirra* when I married; it went by Mount Robalo', and I saw it. It was in a tree. So I started throwing stones. It flew away and came back again. My grandfather said that this bird, which used to sing by night, went in search for water, which didn't exist before. So this bird flew at night scattering *shhh, shhh, shhh* at all sides. Consequently, there was water in all the streams that came down the mountain. The *sirra* gave us water; water we have until now and forever.

'Mejillones and Mount Robalo are locations on the northern coast of Navarino Island at the Beagle Channel, where these Yahgan stories about birds we recorded.

OWLS AND FOREST INTERIOR BIRDS OF PREY

Yahgan: **Sírra**, Síta, Lufkié
Mapudungun: **Chiwüd**, Coa, Coo, Kouw, Yarquen, Chralchral
Spanish: **Lechuza blanca, Lechuza de campanario**
English: Barn Owl
Scientific: *Tyto alba*
(Tytonidae)

Mapuche Territory ▢
Yahgan Territory ▢
Year-round Resident ⬤
Summer Range ⬤
Inhabits a variety of forest and rural, even urban habitats
Sighting probability: 3

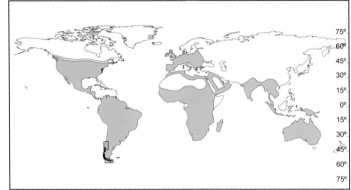

IDENTIFICATION	13-16" (34-40 cm) Unmistakable white facial disc, heart shaped with a dark brown border. Dark eyes and ivory colored beak. Plumage is white to beige on under parts and grayish with brown mottling and small, white spotting on upper parts. Wings and tail are yellowish brown barred with dark brown and its legs have long, white feathers.
HABITAT	This species has the largest world distribution among owls, and has a long history of cohabiting with human settings. In Chile it has been observed in barns, bell towers, abandoned buildings, old trees on farms and plazas, as well as old-growth forests, shrublands and cliffs, from the sea level to the Andean puna, from Arica (17°S) to Navarino Island (55°S) in the Cape Horn Biosphere Reserve.
HABITS	Nocturnal, however in the cold sub-Antarctic forests of Navarino Island it exhibits some diurnal activity. It displays elegant, gliding, slow, and silent flights when hunting over open areas. It nests in a variety of natural or human-made holes, and when chicks are threatened in the nest, they open their wings, lower their heads and move them from side to side while hissing, strongly and constantly.
DIET	It feeds primarily on small vertebrates, particularly rodents, but also birds, bats, reptiles, amphibia and insects.
CONSERVATION	There have been described up to 46 subspecies of *Tyto alba* around the world; 28 of them are generally accepted today. The subspecies *Tyto alba tuidara* (J E Grey, 1829) occupies most habitats in the South American temperate forest biome, and does not present conservation concerns, except for hunting pressures by some farmers and urban dwellers.

Lafkéa
Kill kill
Chuncho
Austral Pygmy Owl

 **CD 1** / Track 13

The *Mapudungun* name for the Austral Pygmy Owl--***kill kill***--repeats the sound of the owl's territorial cry: *kill, kill, kill....* When the Mapuche people of the coast, the *Lafkenche* hear this cry at nightfall, it announces that a man or a woman is leaving their home. More specifically, it reveals the fact that there are problems between the couple. If the woman is abandoned by her husband, she becomes a *kazdtañpe*: a woman who will always be abandoned no matter how many times she marries. In turn, if the man is abandoned, he becomes a *lefmaufe*: a man who will never find a permanent wife.

The squawks of the Pygmy Owl at dusk and night are often raspy and strident and strike fear into the hearts of people because they resemble those of the *tue-tue*, a feared forest spirit. In the former territories of the *Pikunche* people, these nocturnal calls also inspired the *Mapudungun* names *chuchu* and *chonchon*. The *chuchu* inhabits all habitat types in central and southern Chile, and gives origin to names of places, such as Chubunco, in the Valley of Cachapoal river, which means water (*ko*) of the *chuchu*. In the countryside, it is said that the *chonchon* are *winged-heads* creatures, into which witches have transformed themselves. Not surprisingly, then, the *chonchon's* night-time call is believed to announce the presence of a witch:*"kalku akuay" pikey choñchoñ*. To drive it out, the Mapuche people and farmers say prayers, toss salt over the kitchen fire or burn branches of the Winter's Bark Tree or *foye* (*Drimys winteri*).

The Pygmy Owl has other distinctive cries; among these is a repeated short whistle in the form of "ooo-ooo-ooo-ooo-ooo…". When an ornithologist imitates this owl's call in the forest, he or she can attract a variety of birds, which will perch nervously nearby—thus permitting the ornithologist to observe them more closely.

Its distribution is restricted to southern South America. It inhabits a wide range of forest and shrubland habitats from the arid regions of Cordoba and Tarapaca in northern Argentina and Chile to the wet and cold tundra and forest sub-Antarctic ecosystems in Cape Horn, where it receives the Yahgan name of lafkéa. In the austral forests, lafkéa nests in hollow trunks and while it is small and brown-chestnut in coloration, its huge eyes with yellow irises jump out at the observer day and night. It hunts at all hours. Despite being the smallest of the owls in the austral forests, it is a voracious predator of rodents, birds and insects. For this reason, across all its distribution range, the Austral Pygmy Owl constitutes an important ally for country folk to control pest outbreaks.

Yahgan: **Lafkéa,** Lefkóiya, Lö(f)kwīa, Lafkgouia

Mapudungun: **Kill kill,** Chuco, Conchon, Chucho, Chuchu

Spanish: **Chuncho,** Caburé grande

English: Austral Pygmy Owl, Chilean Pygmy Owl,

Scientific: *Glaucidium nanum* (Strigidae)

Mapuche Territory ⬜
Yahgan Territory ⬜
Year-round Resident ⚪
Occasional Visitor ⚫
Summer Range ⚪
Winter Range ⚪

Common in a variety of forested, shrubby, rural, and even urban habitats
Sighting probability: 3

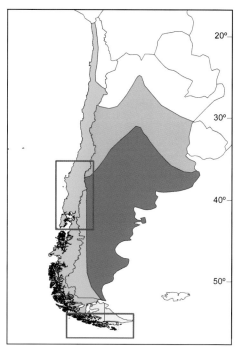

IDENTIFICATION	6.5 - 8" (17-21 cm) The smallest of the owls inhabiting the South American temperate forests. Large head characterized by the notorious yellow irises of its eyes, the white eyebrows, and the two black spots with white borders (resembling eyes) on the nape. Yellow legs. Upper parts grayish or brownish with white spots, while breast and belly with prominent white streaking. The general color of the plumage as well as the number and shape of the transversal bars on the tail are variable, it being possible to distinguish three morphotypes: i) brown plumage with many pale, narrow bars on the tail; ii) gray plumage with only a few pale, broad bars on the tail; iii) rufous plumage, often lacking bars on the tail.
HABITAT	It inhabits from the semiarid shrubby habitats in Atacama (27°S)* to the perhumid sub-Antarctic rainforests in the Cape Horn Archipelago (56°S). Over this extensive latitudinal range it occupies a broad variety of habitats, including old-growth forests, secondary forests and plantations, as well as rural and urban areas, from the sea level to the Andean Cordillera (2,000 m). * *According to some records even from Tarapaca (18°S)*
HABITS	Diurnal, crepuscular, and nocturnal hunter. It is frequently observed during the day perching on exposed branches, or poles of fences. Its presence is also noticed through its unmistakable, loud and long series of whistles.
DIET	Small, but an aggressive and voracious hunter; it captures a broad array of birds, including some species that are larger than itself. It also hunts rodents, reptiles, amphibian, arachnids, insects, and other invertebrates. It often eats only parts of the captured vertebrate; hence, in pigmy owls' territories one can occasionally find intact bird heads or wings dropped on the ground.
CONSERVATION	Abundances of the Austral Pigmy Owl populations are considered to be stationary throughout its distribution range in Chile. Gardens in cities provide habitat for introduced prey birds (passerines, including House Sparrows), as well as for this "human-tolerant" austral owl.

OWLS AND FOREST INTERIOR BIRDS OF PREY

Kulálapij
Pichi kokoriñ
Peuquito
Bicolored Hawk

 **CD 1** / Track 14

To observe a Bicolored Hawk flying among the forest foliage is an unforgettable experience; the elegance and skill of its silent flight is exquisite. Its Mapuche name *pichikokoriñ* refers to the fact that it is a small (*pichi*) hawk (*kokoriñ*). It is about the same size as a Chimango Caracara, but the Bicolored Hawk is distinguished by its dark gray color in the upper face, its blazing yellow irises, and its slim body with its short, rounded wings and long white-tipped tail, striped with six dark bars.

The Bicolored Hawk is a diurnal raptor that inhabits the interior and margins of very diverse forests of the Neotropics, ranging from Central America to the extreme south in Cape Horn. In Chile and Argentina the subspecies *Accipiter bicolor chilensis* inhabits all the *Nothofagus* forest types, and its distribution is restricted to these temperate forests of South America, being isolated from all other subspecies that are found in the neotropical forests: *A. bicolor bicolor* (from Mexico to Bolivia), *A. bicolor fidens* (southern Mexico), *A. bicolor pileatus* (Brazil), and *A. bicolor guttifer* (southern Bolivia and northern Argentina). In addition to the geographic isolation, *A. bicolor chilensis* differ from the other subspecies by its coloration, morphology, and behavior, and for these reasons it is considered to be a separate species, *Accipiter chilensis*, by many ornithologists.

Accipiter bicolor chilensis has been classified as a rare species due to its low population densities over much of its range. However, it is seen relatively frequently in the sub-Antarctic forests of the Magellanic Archipelago in the Yahgan territory south of Tierra del Fuego. Its Yahgan name, *kulálapij*, expresses its aggressive character, identifying it as an angry (*kulala*) bird (*pij*). Indeed, *kulálapij* is a skilled hunter inside the forest, able to catch flying birds as well as those perching or moving among the foliage. Its diet includes forest birds of small (e.g., Thorn-Tailed Rayaditos and House Wrens) and medium sizes (e.g., Eared Doves and Austral Thrushes), and occasionally birds of larger size outside the forests, such freshwater ducks, and domesticated chicken and quails. It also hunts rodents and insects among the foliage and on the ground. In addition, in the canopy of tall trees, pairs of *kulálapij* construct their nests which form a platform about 60 cm wide over high, forked branches radiating from the trunk. They are resident, and use the same nest for several years. In short, *kulálapij* is a bird inextricably linked to the forests, or *ashuna*, through its diet, reproduction, and territorial habits.

Yahgan: **Kulálapij**, Chuhchul

Mapudungun: **Pichi kokoriñ**

Spanish: **Peuquito**, Esparvero variado, Cazapollo

English: Bicolored Hawk, Chilean Hawk

Scientific: *Accipiter bicolor chilensis* (Accipitridae)

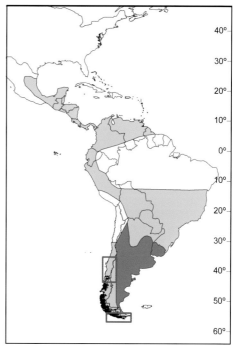

Mapuche Territory

Yahgan Territory

Year-round Resident, but some migrate north during winter

Occasional Visitor

In a variety of forest types
Sighting probability: 1

IDENTIFICATION	14-17.5" (36-44 cm); Male 14-15" (36-39 cm), Female 15-17.5" (39-44 cm) Slim raptor with bright yellow iris and eyelid, short wings and long dark brownish gray tail with five or six darker transverse bands. Adults have plumage dark blackish brown in their upper parts, and pale white grayish or brownish with dark ash gray transverse bands on their chest and abdomen. Juveniles are distinguished by their light green irises, gray eyelids, and the whitish plumage of their chests with brownish streaking.
HABITAT	*Accipiter bicolor chilensis* inhabits a variety of forest types, including forest edges, and adjacent open lands. It occasionally visits towns and cities.
HABITS	It prefers dense forests interspersed with open areas that allow a good habitat for hunting and shelter. Thanks to its slender body, long tail and short rounded wings, it has a fast and direct flight that has allowed it to become one of the most forest specialized predators. Its call consists of a sharp series of *squeaking* notes.
DIET	It feeds on birds preferably, but it will also capture rodents, reptiles, and insects.
CONSERVATION	In Chile some authors consider the subspecies *Accipiter bicolor chilensis* a species (*A. chilensis*); under both taxonomic levels it has been classified as vulnerable due to its declining population trend. **V**

OWLS AND FOREST INTERIOR BIRDS OF PREY

WETLAND BIRDS,
ASSOCIATED WITH RIPARIAN, COASTAL OR PRAIRIE HABITATS

Chéketej
Challwafe üñüm
Martín pescador
Ringed Kingfisher

 **CD 1** / Track 15

Ceryle torquata, formerly classified as *Megaceryle* (great = Gk. *megas*; kingfisher= Gk. *ceryle*) *torquata*, is the largest South American kingfisher, and it is the only one that reaches subpolar latitudes. Its distribution spans from Texas and Arizona in southern United States to Cape Horn in southern South America, where it receives the Yahgan name of *chéketej*.

It is a conspicuously colored species with an elegant white collar and a blue crest, especially marked in the male. It possesses a long, strong beak that permits it to catch fish in rivers, lakes, channels and fjords of the extreme south. It is frequently observed perched on branches or rocks that overhang rivers or the shoreline. On these, the Ringed Kingfisher waits for the appearance of marine and freshwater fish, crustaceans and larvae that it hunts on the surface of the water or by plunging itself into it. When it notices danger, it sweeps back and forth in the air, emitting its strong, repetitive calls *kekereke- kekereke- kekereke*.

It nests in holes in the sides of riverbanks and hills. When she was young, Yahgan Grandmother Cristina Calderón enjoyed watching the Ringed Kingfisher, or *chéketej*, coming in and out of their nests on cliffs along the Murray Channel on Hoste Island. Some old Yahgan stories related *chéketej* with another bird species that uses forests and waters, the Night Heron *(Nyctiocorax nyctiocorax)* or *huajatanu*. Both the Ringed Kingfisher and the night heron sleep and nest in trees, but fish in rivers or along shorelines. In ancestral times when *chéketej* and *huajatanu* were still humans, they were lovers. For a long time, whenever *huajatanu* left to fish in the canoe, *chéketej* helped him to catch great quantities of fish, until one day the *huajatanu's* husband surprised the lovers and impaled them with his *sírsa* or sea urchin harpoon; it was then that both transformed themselves and flew away as birds.

Grandmother Cristina was enchanted by the colors of the male and the female when they flew as a pair into the waters to capture fish. This "enchantment" is also captured by the Mapuche ornithological view of the Ringed Kingfisher or *challwafeüñüm*, which perceives this species as a "half tree-half water" fisher (*challwafe*) bird (*üñüm*). The *challwafeüñüm* attracts the fishes with its attractive, iridescent colors that it displays in its flights skimming over the water, plunging itself into the water from sunrise

WETLAND BIRDS

to sunset. Through the hypnosis that *challwafeüñüm* exercises over the fishes, an ecological flow from the beings of the water to the beings of the air is generated; the bird's beauty, itself, forges the links in this food chain. Here, the insight of the traditional Mapuche ecological knowledge has essential similarities with contemporary ecological concepts of food webs, and the importance of the role played by birds such as the Ringed Kingfisher , which transport nutrients between aquatic and terrestrial ecosystems, contributing to the fertility of the land. In words of Mapuche poet, Lorenzo Aillapan:

MAPUCHE STORY **CD 1**/ Track 54

The kingfisher or *challwafeüñüm* attracts fish with its flashy colors, which it displays in its flights which skim the water. This hypnosis, that the kingfisher exercises over the fish to consume them, generates the ecological flow between the beings of the water and the beings of the air by means of the food chain.

MAPUDUNGUN VERSION

Fachi CHALLWAFEÜÑÜM feytañi allangechi fill wiriñ pichuñ wüñowitra mütrümkefi pu CHALLWA mülelu leufü meu üpüm ketu amuy wenteko. Fey tachi CHALLWAFEÜÑÜM mütrüm adtukey reke feyti illun challwa duamfel nümün epuñpüle üllutuwün fill kulliñ mülekelu tuwe ka kürüf meu fey ñi ipangentuaal.

Yahgan: **Chéketej,** Šégetex, Chakatakh

Mapudungun: **Challwafe üñüm,** Queto, Quete, Kedküchan

Spanish: **Martín pescador**

English: Ringed Kingfisher

Scientific: *Ceryle torquata* (Alcedinidae)

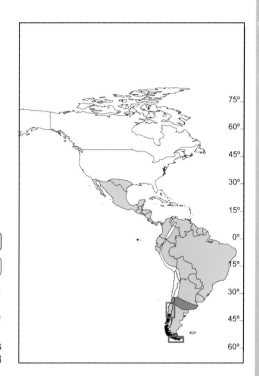

Mapuche Territory ☐

Yahgan Territory ☐

Year-round Resident ⬤

Occasional Visitor ⬤

Common in forested coastal and riparian habitats
Sighting probability: 3

IDENTIFICATION	14-17" (36-44 cm) Deep blue or blue-grayish plumage with white markings, a shaggy crest and a broad white collar around the neck. The male stands out for its rufous chest and belly, and the female for its blue-grayish chest separated by a white band from its rufous belly. It has a strong, long, straight beak of dark gray color.
HABITAT	In Chile, the subspecies *Ceryle torquata stellata* inhabits the forests, woodlands, shurblands, and cliffs around lakes and rivers, as well as fjords, marine channels and sheltered bays, from Colchagua (35°S) to the Cape Horn archipelago (55°S).
HABITS	They perch for long periods on prominent branches of trees, rocks, bridge rails or cables near water. From these suitable watchpoints, pairs or single birds fly over the water, frequently emmitting their high pitch, percussive vocalizations, before plunging into the water catch a fish. They nest on coastal cliffs where they build deep galleries.
DIET	As do all the other kingfisher species that inhabit the Americas, *Ceryle torquata* belongs to the Cerylidae family, which specializes in feeding on fish.
CONSERVATION	The Ringed Kingfisher posses a wide geographical distribution in the Americas, and three subspecies have been identified. The biome of the South American temperate forests is inhabited by the subspecies *Ceryle torquata stellata*. Both the species and the subspecies are listed as "Least Concern" taxa by Birdlife International, and IUCN.

Shakóa
Porotu
Becasina
Common Snipe

(•) **CD 1** / Track 16

The *Mapudungun* name *porotu* is onomatopoeic with the sound that the Common Snipe male emits during the reproductive period. At dusk, during the spring, the male rises into the air and then falls back to the ground, making his tail feathers vibrate and generate a strange hum: "*porotu, porotu, porotu, porotu, porotu.*" In the spring, shrimp also make their appearance. Due to this synchrony, the strange twilight sound of the *porotu* announces harvest time for the *madeu* or shrimp, a favorite food of the *Lafkenche* people. Because of its talent for extracting the *madeu* with its long beak from the caves they make in the mud, the *porotu* is also known as the shrimping-bird in the *Lafkenche* region on the coast of Temuco.

The South American Snipe is characterized by a long bill, which they use to eat invertebrates buried in the mud. Snipes live in a variety of aquatic and riparian environments, especially the inundated wetlands and rush thickets adjacent to the forests of southern Chile and Argentina. The South American Snipe was considered until recently to be a subspecies (*Gallinago gallinago paraguaiae*), but is identified as a separate species, *Gallinago paraguaiae*, and the South American temperate forest biome is inhabited by the subspecies *Gallinago paraguaiae magellanica*.

Darker and more robust than the South American Snipe, the Fuegian Snipe, *Gallinago stricklandii* (35 cm), also lives in southern South America. It is a resident of the territory south of Tierra del Fuego, where it receives the Yahgan name *shakóa*. The Fuegian Snipe is inconspicuous because its plumage mimics the colors and patterns of the grasses and rushes, and when one approaches them they remain immobile. However, their presence is noticed at dusk during the spring, when the male elevates into the air, making his loud courtship humming sound.

Among the *Yahgan* it is not permitted to imitate the sound of the snipe or *shakóa*. She or he who imitates the sound of the *shakóa* will awake with her or his toes cut by blades of *ushkulampi* or rush

WETLAND BIRDS

(*Marsippospermum grandiflorum*) that grow in the bogs. In fact, the grandmothers Úrsula and Cristina Calderón remember that when they were little girls one day they awoke with the big toe of each foot cut because of imitating the *shakóa*, even when they had covered their feet well. They tell that:

YAHGAN STORY

 CD 2/ Track 4

My father and my grandparents and my mother said that one should not imitate the snipe or *shakóa* when it is flying during the night. "If you imitate the bird, they will hurt you. They will make a cut in your big toe with grass or *kuruk*," they said.

As we were naughty and we wanted to test if the things were certain or not we decided to try it. We were three little girls. So we wrapped our feet with a very thick cloth and with a sock over it, and began to imitate the bird at night. The next day we woke up with a lot of pain in our feet, because the story of the snipe was true. We were certainly naughty girls, but after that we never did it again because we knew that it was certain what the grandparents told us.

Yahgan: **Shakóa**, Tsakaoa

Mapudungun: **Porotu,** Carcaren, Cadcadeñ, Kaikayem, Kedkedeñ

Spanish: **Becasina**, **Porotera,** Becacina, Becacina común

English: Common Snipe, South American Snipe

Scientific: *Gallinago paraguaiae*

               (Scolopacidae)

WETLAND BIRDS

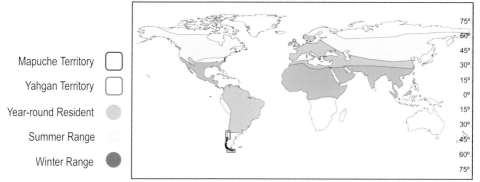

- Mapuche Territory
- Yahgan Territory
- Year-round Resident
- Summer Range
- Winter Range

Resident northern populations; extreme southern populations migrate north during the winter
In a variety of wetland habitats. Until recently, *Gallinago paraguaiae* was considered as one of
the subspecies of *Gallinago gallinago* which had a wide world distribution, depicted in the map.
Sighting probability: 2

IDENTIFICATION	9-12" (23-32 cm) Characteristic longitudinal white bands forming a type a "V" on the back. Upper parts yellowish brown mottled with black. White belly. Head with two white lines running from the base of the beak toward the nape, one above the eye, another below the eye. Long (69 mm), straight, straight dark olivaceous bill with black tip. Olive-yellow legs. Distinctive rufous brown band at the tip of the tail.
HABITAT	Wetlands, including salt marshes, wet grassy steppes, rush thickets, peat bogs and boggy rivers. The subspecies *Gallinago paraguaiae magellanica* inhabits from the sea level to 2,000 m in the Andean valleys; from Mendoza (Argentina) and the Copiapo valley (27°S, Chile) to Horn Island (56°S) at the southern end of the Americas.
HABITS	It performs a characteristic aerial courtship display during twilight, flying high in circles and then taking shallow dives, producing a distinctive sound with its tail. When disturbed it emits a loud, rasping *tssk* call, and takes off in zigzagging flight, landing again at a short distance. Mimetic, hides in wetland vegetation; hence, they are easier to hear than to see. Builds nest with rushes and grass, where they lay two eggs toward the end of winter.
DIET	Feeds on adult and larval insects, earthworms and other invertebrates, probing and picking in the soft mud with its long beak.
CONSERVATION	*Gallinago paraguaiae* classified as a species of Least Concern by IUCN; the subspecies *G. p. magellanica* does not present conservation concern either. However, the Fuegian snipe (*Gallinago stricklandii*) is a poorly known species, which has small populations that may be declining in some areas due to wetland habitat degradation; hence, it is classified as Near Threatened by IUCN. **NT**

WETLAND BIRDS

Waior
Pideñ
Pidén
Plumbeous Rail

 **CD 1** / Track 17

The strident call of the Plumbeous Rail characterizes the swampy landscapes in the temperate forest biome of southern Chile and Argentina, and gives rise to the *Mapudungun* name *pideñ*, onomatopoeic with its loud call *pidreeen, pidreeen, pidreeen*. This noisy bird is found in wetlands from southern Brasil, Paraguay, Bolivia, Peru, and Ecuador to the southern end of South America, where it receives the Yahgan name of *waior*. In the sub-Antarctic forests and tundra the *waior* inhabits a variety of wetlands, including river banks, and more recently ponds created by beavers introduced from Canada. In these habitats, *waior* moves with skill among the rushes (*Marsippospermum grandiflorum*) or *ushkulampi*, placing one leg before the other in such a way that tracks are left in a single line. Further north it also inhabits and moves among the Southern Cat-tails (*Typha spp.*) or *trome* of the wetlands of farms, coastal and Andean habitats of the Mapuche territory.

Its legs and eyes are an intense red, while its body is somberly-muted with a blackish-blue chest and an olive-brown back. It has a long, multicolored beak (blood red at the base, bluish in the middle and a greenish tint at the tip) which it uses to feed upon a large variety of invertebrates and vegetable material. In order to be heard in the dense vegetation, the Plumbeous Rails emit a strong, prolonged series of cries, especially at dusk when they listen to the answers of different members of the population.

For the Mapuche people, the *pideñ* is a bird that forecasts the weather with its calls at the end of the day. On Chiloé Island, the *Williche* and farmer communities predict that if the rails sing in chorus at sunset, the following day will have good weather. The *Lafkenche* say that if the *pideñ* projects its calls towards the north, there will be a torrential downpour; if they call towards the south there will be good weather. As the *Lafkenche* poet Lorenzo Aillapan recites:

MAPUCHE STORY

 CD 1/ Track 55

The strident call of the *pideñ*, *cotuta*, or *gallineta* characterizes the swampy landscapes in the south of Chile and Argentina. As the bilingual poem *Pideñkawun* says:

The resounding cry of the Plumbeous Rail
Gathered together they sing, and from the matorral they depart,
Announcing bad weather and rain.
In the dense river-cane they hide,
Wading birds with red legs,
Beaks similar to the Oystercatcher.
¡Piiiir with piiiir with fiiiir with fiiiir with
piiiir with piiiir with fiiiir with fiiiir with!

MAPUDUNGUN VERSION

Tachi waken ülkantun pidén, cotuta achawall pingey kimfaluwi rumel focholen adlelfün willipüle Chile/mapu ka Argentina/mapu. Feypiley tachi epu rume dungun koyautun. Pideñkawün meu:

Pideñ tañi fillke kümüllün wakeñnün
Fentren üñüm re kefafan tripapay walfe mew
Rüf küftukun mawünay pileyngün
Cheu ñi mülen trongen külantu ellkaupuyngün
Epu kelü namun tuley rofi küpal üñüm
Taiñ komilwe wün pilpilen üñüm femngey.
¡Piiiir with piiiir with fiiiir with fiiiir with
piiiir with piiiir with fiiiir with fiiiir with!

WETLAND BIRDS

Yahgan: **Waior,** Heuhs, Wéásh

Mapudungun: **Pideñ,** Piden, Cotuta

Spanish: **Pidén, Gallineta común,** Cotuta

English: Plumbeous Rail

Scientific: *Pardirallus sanguinolentus* (Rallidae)

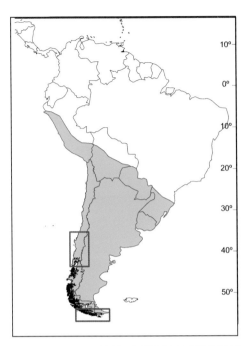

Mapuche Territory ☐

Yahgan Territory ☐

Year-round Resident

Common in forested coastal and riparian habitats
Sighting probability: 3

IDENTIFICATION	11-16" (27-40 cm) Dark olive-brownish plumage on the upper parts of the body, while the face, chest and lower parts have a paler ashy-blueish color. Long, slightly curved, green yellowish bill, with bluish base at the upper mandible, and red spots at the base of both mandibles. Red iris and legs. Immature birds are completely brown, and have gray iris and legs.
HABITAT	Inhabits all types of wetlands with dense vegetation, including riparian areas, coastal grasslands, marshes, beaver ponds, tundra and estuaries; from sea level to the high Andes (3,000 m).
HABITS	Their calls form characteristic choruses at sunset, when individuals answer to each other. They are more active at this time of the day, and at night when they search for food in more open areas. During the day they are secretive and shy, and rarely leave the dense vegetation; hence, they are easier to hear than to see. They nest in these swampy environments where they build simple grassy platforms laying 4 to 7 eggs. They are agile walkers, but flight poorly, and swim only when they escape or cross from one shore to the other.
DIET	It feeds on variety of invertebrates, including insects larvae and earthworms, and occasionally some parts of plants.
CONSERVATION	Due to its generalist habits in the use of wetlands areas, and wide distribution in South America, *Pardirallus sanguinolentus* is classified as a species of Least Concern by Birdlife International and IUCN. It is important to note, however, that of the six subspecies that have been identified, only two inhabit the South American temperate forests biome: *Pardirallus sanguinolentus landbecki* (from Neuquen to Santa Cruz in Argentina, and from Atacama to Aysen in Chile), and *P. s. landbecki* found in Tierra del Fuego and the Cape Horn archipelago. In the sub-Antarctic ecoregion these subspecies are sympatric with the Austral Rail (*Pardirallus antarcticus*), one of the least known species of the Neotropics. This very rare species is classified as Vulnerable by IUCN. **V**

WETLAND BIRDS

Tulára-táchij
Chiuchiu
Churrete acanelado
Bar Winged Cinclodes

(•) **CD 1** / Track 19

Besides the Bar Winged Cinclodes, there are several other species of Cinclodes, which inhabit the South American temperate forest biome, and have similar calls. Consequently, they all receive the same *Mapudungun* onomatopoeic name *chiuchiu*. Among these species, the Bar Winged Cinclodes or Dusty Cinclodes has the broadest geographical distribution, and is the most generalist in the use of habitats. At the the southern extreme, on Navarino Island and the Cape Horn Archipelago it is the only Cinclodes that follows the streams, or *yaha-huen*, uphill, inhabiting the slopes and summits of the mountains. For this reason, its Yahgan name is *tulára-táchij*, which describes this bird as a Cinclodes (*táchij*) that principally inhabits the mountains (*tulára*). Interestingly, the coloration of *tulára-táchij* is brownish or cinnamon, which makes this bird or *pij* more mimetic to the *tulára* or mountainous environments.

As with the other species of Cinclodes in the sub-Antarctic forest ecoregion, it is not rare to find the Bar Winged Cinclodes marauding in the pools and wet areas in the interior of the evergreen forests, even in the deep winter snow or *panaja*. But in the Magellanic Evergreen Beech forests and other coastal ecosystems of the Yahgan territory in Cape Horn, the presence of another species of Cinclodes might surprise us: the Blakish Cinclodes (*C. antarcticus),* which is the only species of Cinclodes without a wing bar. This species is endemic to the sub-Antarctic ecoregion, and its preferred habitats are the tussock grasslands of the exposed islands and coastlines it shares with colonies of sea birds and sea lions.

WETLAND BIRDS

Yahgan:	**Tulára-táchij,** Toularatachigh, Tatchigh
Mapudungun:	**Chiuchiu**, Chiuchihuen
Spanish:	**Churrete acanelado, Remolinera común**
English:	Bar Winged Cinclodes, Dusty Cinclodes
Scientific:	*Cinclodes fuscus* (Furnariidae)

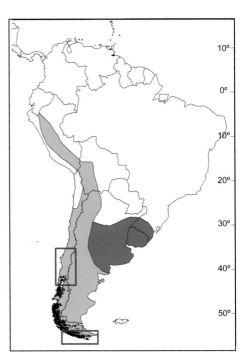

Mapuche Territory ☐

Yahgan Territory ☐

Year-round Resident

Occasional Visitor ⬤

Along streams or water bodies on a variety of habitat types
Sighting probability: 3

IDENTIFICATION	6-7" (16–18 cm) Upper parts ashy-brown, with pale grayish-brown lower parts. Whitish throat with light brown streaking. Conspicuous whitish supercilium, and characteristic redish brown wing-bar visible in flight. Bill is short, straight, and pointed.
HABITAT	In a variety of wetland and riparian habitats, along streams on mountainous slopes, rivers, ponds, or lakes; from the sea level to open the high-Andes (>4,000 m).
HABITS	During the breeding season it emits a characteristic loud vocalization. Largely terrestrial, it runs and hops over the ground, and from time to time perches in shrubs. It nests mostly on montainious slopes, on stony slopes or cliffs.
DIET	It feeds on insects and other small invertebrates.
CONSERVATION	*Cinclodes fuscus* is the most widespread species in the genus, and includes nine subspecies. *Cinclodes fuscus fuscus* is present in the temperate forests biome; both the species and subspecies are classified as taxa of Least Concern by IUCN.

Táchij
Chiuchiu
Churrete
Dark-Bellied Cinclodes

(•) **CD 1** / Track 20

All the Cinclodes species emit similar vocalizations, and receive the onomatopeyic *Mapudungun* name *chiuchiu*, which derives from the bird's short calls *chrrr-chrrr*. All the Cinclodes also have similar behaviors, flying from rock to rock, moving continuously without stopping. They only rest for brief instances to look in every direction while they bob their tail and emit their short call, *chrrr-chrrr* or *chiuchiu*. From this behavior comes the Argentinean names *remolinera* or *meneacola* (tail-bobber). In contrast, its Chilean name, churrete, is derived from the fact that these birds have a very active digestive system, and leave behind patches of feces that stain the rocks on the shore. In Chilean slang "to be *churrete*" means to have diarrhea. In the sub-Antarctic archipelago it receives the Yahgan name *táchij*, and fishermen and sailors today call this bird *piloto*, or captain. This name is derived from its characteristic white markings around the eyes that resemble navigator's goggles, and the fact that *táchij* is continuously moving or "navigating" along the shorelines of the austral channels and fjords.

As we approach the shorelines of rivers, lakes, as well as fjords, channels and beaches in the austral temperate forest region, the *chiuchiu* surprises us with its characteristic short vocalization, *prrr, prrr, prrr*, as it skips from rock to rock. Grayish-brown, its coloration mimics the rocks and stones where it hops about in search of all types of crustaceans, mollusks and invertebrates. The only showy parts of this bird are its eyebrow that extends to the nape of the neck, its pale throat, and its song.

During the reproductive season it is common to hear the *chiuchiu* sing its melodious, sharp, strong, prolonged, and trilled song, while it beats its wings, perched on an overhanging branch or rock. *Chiuchiu* nests in natural hollows among rocks or fallen logs, as well as in buildings adjacent to water bodies or streams, and they commonly nest below bridges.

Yahgan: **Táchij**

Mapudungun: **Chiuchiu**, Churrete, Churreta, Churete, Chiuchihuen, Thureu, Huechuthureu

Spanish: **Churrete, Remolinera araucana,** Remolinera común, Meneacola, Piloto

English: Dark-Bellied Cinclodes

Scientific: *Cinclodes patagonicus* (Furnariidae)

Mapuche Territory ☐

Yahgan Territory ☐

Year-round Resident ⬤

Along streams, water bodies and coastal habitats
Sighting probability: 4

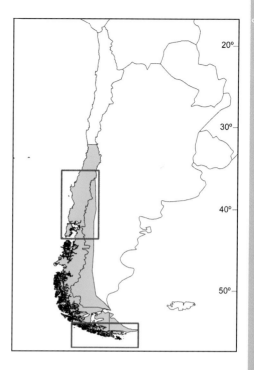

IDENTIFICATION	7-9" (18-22 cm) Dark brown head with a distinctive long and wide white eye line, which extends to the neck. Upperparts, wings, back and tail dark brown. Underparts are smoky-gray with streaks that range from brown to white in color.
HABITAT	A variety of riparian and coastal habitats, including saline or freshwater environments, as well as around human settlements built near freshwater or coastal areas.
HABITS	It is common to observe it along river edges in forested areas. It is very active and confident around humans. They prefer to nest in holes or cracks in cliffs at the sides of rivers, places with soft ground that allow them to excavate, and also in abandoned houses or under bridges. During the breeding season it has a characteristic loud and long song. While it forages and moves along water bodies and streams it emits shorter, continuous vocalizations.
DIET	Voracious feeder, it eats insects, mollusks and other small invertebrates.
CONSERVATION	It is classified as a species of Least Concern by IUCN.

Lásij
Pillmaykeñ
Golondrina chilena
Chilean Swallow

(•) **CD 1** / Track 21

The swallows of the forests of southern Chile and Argentina include several migratory species that arrive in spring and leave in fall. Among them, the Chilean Swallow and the Blue-and-White Swallow are the most common. Both species receive the *Mapudungun* name *pillmaykeñ*, which means spirit (*püllü*) that travels (*may;* the suffix *keñ* makes the name a noun), and their arrival announces the spring. The Yahgan language also assigns the same name, *lasij*, to both the Chilean Swallow and the Blue-and-White Swallow.

Sometimes the Chilean Swallow was called *lasijkipa* in Yahgan because it was considered one of the wives (*kipa* = woman) of the Wren or *chílij (Troglodytes aedon)*, another small bird with migratory habits. This pairing was probably inspired by the observation that, in the extreme south, *lasijkipa* and *chílij* both nest in cavities of banks or hollows in standing dead or live trees. These species take advantage of naturally occurring cavities and those made and abandoned by woodpeckers.

The *Williche* communities of Chanquín and Huentemó south of Chiloé National Park carefully observe the type of flight of *pillmaykeñ* in order to forecast the weather: if the *pillmaykeñ* fly high they announce good weather; if their flight is low to the ground, the weather will be bad. Swallows play an important role in the control of pests in the forests and agricultural fields, given that they fly tirelessly hunting for insects. That is why they are frequently seen flying artfully and doing pirouettes; their aerial dance allows them to consume millions of insects.

The Chilean Swallow is identified by its elegant bluish-black color on its back which is interrupted by a white stripe before the tail. Its chest is also white. In the Mapuche territory, these swallows provide the source of the name of the beautiful Pilmaiquén River (*Pillmaykeñlewfü*) with its famous waterfall. The Chilean Swallow nests in the roofs of houses in the cities and countryside of Chile, its arrival announces springtime and romances. That is why many girls are named *Pillmaykeñ* as told in the following story recorded with the poet Lorenzo Aillapan.

MAPUCHE STORY

 CD 1/ Track 56

The Mapuche name for the swallow is *pillmaykeñ*, which means spirit that travels, flying about the earth. Its arrival announces the spring, that is, the beginning of the period of courtship. Hence, the swallow excites the young people who frequently are given the name *pillmaykeñ*. During the 1990's, the tradition of giving sons and daughters bird names has revived in the Mapuche communities. Today, the name *pillmaykeñ* is as popular, or more popular, among little girls as the Christian name María.

MAPUDUNGUN VERSION

Tachi üytukun pillmaykeñ tripalu püllü kañpüle ka fey akutulu weflu pewü, müleyüm dakelüwün duamtu, fey pu wekeche ülcha deuma pillmaykeñ üytuy. Fey pu lofche, rupalen tripantü 1990, pu kurewen che femngechi pu fill üñüm üy tukutuy ñi pu fotüm ka pu ñawe. Femngechi, lofmeu kiñeke doy allmantuy tukual pillmaykeñ üy chumngechi mülefuyta wera fentren ka feyentun pu malen "María" pingekefulu.

Yahgan:	**Lásij,** Lášix, Lášixkipa, Lazikh
Mapudungun:	**Pillmaykeñ,** Wüshükon, Pillmayken
Spanish:	**Golondrina chilena,** Golondrina patagónica
English:	Chilean Swallow
Scientific:	*Tachycineta meyeni* (Hirundinidae)

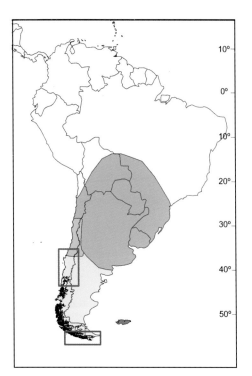

Mapuche Territory ☐

Yahgan Territory ☐

Year-round Resident ●

Occasional Visitor ●

Summer Range ●

Winter Range ●

Resident northern populations; southern populations migrate north during the winter
In a variety of habitat types
Sighting probability: 4

IDENTIFICATION	4-5.5" (11-14 cm) Small swallow with glossy blue-black plumage on the upperparts, and white on all underparts. Distinctive white rump.
HABITAT	A variety of habitats in the vicinity of surface water, such as streams, rivers, ponds, or lakes. It also lives in towns and cities, as well as open grasslands. Found from the sea level to the high-Andean (2,500 m).
HABITS	Forms large flocks, emitting a variety of gurgling vocalizations. Skilled flyer, catching insects during flight, and also over water surface on lakes and intertidal ponds. They nest in holes on cliffs and trees, as well as in the cracks of the roofs of houses.
DIET	It feeds on insects caught during flight.
CONSERVATION	It is classified as a species of Least Concern by IUCN.

Lásij
Pillmaykeñ
Golondrina de dorso negro
Blue-and-White Swallow

 CD 1 / Track 21

The Blue-and-White Swallow is another common swallow species in the forests of southern Chile and Argentina. As indicated by its name, this swallow has no white band on its back; it is also more selective with respect to its preference for aquatic habitats than the Chilean Swallow. The Blue-and-White Swallow abounds near bodies of water where mosquitoes and other insects thrive. As they fly about in curling masses, their polyphony of voices suggests, as indicated by the *Lafkenche* poet Lorenzo Aillapan, that *pillmayken* engages in a "birdly conversation."

The Blue-and-White Swallow is migratory and covers a wide range of territory from Mexico to Navarino Island, Chile. In the fall, it migrates north and returns each spring to the austral forests where it is almost always seen chasing insects over the rivers, lakes and even fjords. The Chilean Swallow's reputation for being a good hunter might be related to an ancient Yahgan narrative.

In ancestral times when birds were still humans, the swallow or *lasij* was a small hero who jumped into the mouth of a giant dolphin, or *wéoina*, which swallowed it. Once it was inside the enormous dolphin, the little *lasij* used a sharp knife to cut the internal organs of the marine mammal. *Lasij* needed a great deal of time to kill *wéoina*. Only when *lasij* began to hear petrels, or *tawísiwua*, and albatrosses, or *karpo*, flying outside did the hurt dolphin swim exhausted towards the beach. There *wéoina* died and provided abundant meat and blubber for all the Yahgans. Thanks to *lasij*, the Yahgans had food that permitted them to hold the celebration of the *kina*, an important initiation rite of young boys, which had not been possible in a long time.

113

WETLAND BIRDS

Yahgan:	**Lásij**, Lášix, Lášixkipa, Lāsix, Lazikh
Mapudungun:	**Pillmaykeñ,** Wüshükon, Pillmayken
Spanish:	**Golondrina de dorso negro**, Golondrina barranquera
English:	Blue-and-White Swallow
Scientific:	*Pygochelidon cyanoleuca* (Hirundinidae)

Mapuche Territory ☐

Yahgan Territory ☐

Year-round Resident ●

Summer Range ●

Resident northern populations; southern populations migrate north during the winter
Common in a variety of habitat types
Sighting probability: 4

IDENTIFICATION	5" (11-12 cm) It has dark blue upperparts and white underparts. Similar to the Chilean Swallow, but lacks the white rump; also, its underwings and the undersurface of its short, forked tail are blackish.
HABITAT	In contrast to the Chilean Swallow, the Blue-and-White Swallow is rarely found near human settlements. It inhabits steppes, shrubby areas, forest edges, cliffs, and open areas near watercourses; from sea level to the high Andes (4,000 m).
HABITS	Generally in flocks, including mixed flocks with the Chilean Swallow. Its flight is typically fluttery; it is usually seen catching insects during flight. They continuously emit a *dzzzhreeee*, buzzing vocalization. This swallow frequently perches on branches of trees. They build their nests in deep caves or cracks, using roots, grasses, and feathers.
DIET	It feeds on insects that it catches during flight.
CONSERVATION	It has a wide geographic distribution, and it is classified as a species of Least Concern by IUCN.

Lejuwa
Raki
Bandurria
Buff-Necked Ibis

CD 1 / Track 22

The loud metallic calls of the Buff-Necked Ibises distinguish the temperate forests of southern South America. Their strong vocalizations lead to its onomatopoeic *Mapudungun* name *raki*. The calls of the *raki* are used by the *Williche* communities to forecast the weather. On Chiloé Island it is said that when Buff-Necked Ibises call half way up the mountain, they announce good weather. In contrast, when they shout with their wings open, perched on the trees or crow in chorus, they announce bad weather. When they cry at the moment of taking flight, then the wind will start blowing from the south, and the weather will get better.

Moving in flocks, the ibises clean the farmlands of insects that could be harmful to the plantings. In the fields and other open habitats it is common to observe numerous ibises scratching with their long beaks in search for earthworms, larvae, snails and other invertebrates. To sleep and nest, they use the canopy or branches of tall trees, where they come together in colonies. At the end of fall, the populations migrate to the north along the length of the Andes, returning to the south to announce spring. The Yahgan culture also associates the migratory movements of the Buff-Necked Ibis or *lejuwa* with marked seasonal or climatic changes, as is appreciated in the following history narrated by Yahgan Grandmother Cristina Calderón.

WETLAND BIRDS

YAHGAN STORY

 CD 2/ Track 3

In the times of the ancestors, one day when spring arrived, a *yamana** leaned out of his *akar* or hut and saw a Buff-Necked Ibis or *lejuwa* flying in the sky. The *yamana* was so happy that he yelled to the other members of the community, "An ibis is flying over our *akar*. Look!" Immediately the others came out of their huts, shouting loudly, "The spring has come, the ibises are already in their return flight." They jumped for joy, and they jumped hard.

Lejuwa is, however, a very delicate and sensitive woman and likes to be treated with special deference. Upon hearing the noise of the shouts, she became furious, and profoundly offended, causing a strong snow storm. It snowed incessantly; it was freezing cold; all the land was covered with ice; and all the waters froze. Many people died because they could not navigate in their canoes to search for food. Neither could they leave from their *akars* to look for firewood because all was covered with snow. More and more people continued to die. Only after a long time did it stop snowing, and the sun begin to shine.

The sun warmed so much that it begin to melt the ice and snow that had completely covered the land, and consequently floods began to flow towards the channels and the oceans. The ice that covered the wide and narrow channels broke apart and melted, and the Yahgans could finally go to the shores and navigate in their canoes to collect food. On the big mountainsides and deep valleys, however, the ice was so thick that not even this sun could melt it. Even today, glaciers can be observed coming down to the sea, that remind us of the severe frost and snowstorm that *lejuwa*, who is a very delicate and sensitive woman*,* provoked in those times. Since then, the Yahgans treat the Buff-Necked Ibis or *lejuwa* with much respect, and when they come near the houses or *akars*, the people keep quiet, particularly the little children who are not allowed to imitate them or to cry.

*Yamana is an ambiguous term because this is the name of the ethnic group and refers to a Yahgan man. In this case it refers to the second meaning.

Yahgan: **Lejuwa,** Léxuwa

Mapudungun: **Raki,** Raqui, Raquin, Rakiñ

Spanish: **Bandurria,** Bandurria austral, Bandurria común, Bandurria baya

English: **Buff-Necked Ibis,** Cinnamon- Necked Ibis

Scientific: *Theristicus melanopis* (Threskiornithidae)

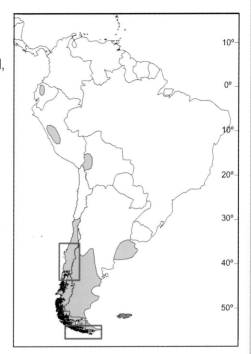

Mapuche Territory ☐

Yahgan Territory ☐

Year-round Resident ⬤

Occasional Visitor ⬤

Resident northern populations; southern populations migrate north during the winter
In a variety of forest and open habitat types
Sighting probability: 4

IDENTIFICATION	29-30" (73-76 cm) Ibis which has a long curved, black bill. Its head, neck, chest and upper belly are orangish-beige, and its lower belly is black. It has a gray collar accross the upper breast. The upper parts are also gray. Dark pink legs.
HABITAT	They perch and nest in a variety of forest types, and feed on wetlands, prairies, marshes, and steppes near the forests or woodlands. From the sea level to the Andes (3,000 m). It is also a frequent visitor of farmlands.
HABITS	Migratory, its arrival to the sub-Antarctic region announces the beginning of spring; they migrate to the north at the beginning of the autumn. Higly gregarious, flies and forages in flocks, and forms large breeding colonies. It feeds over soft mud and soft-soiled grasslands. As predators approach them, they exert strong loud bugling calls. The nest is usually placed over a tree or on a cliff. Females usually lay two to four eggs in a platform nest made of twigs and branches.
DIET	Feed mainly on insects, snails, and other invertebrates, frogs and tadpoles, and occasionally on reptiles that can be found in soft soils.
CONSERVATION	It is classified as a species of Least Concern by IUCN.

Pílit
Tregül
Queltehue
Southern Lapwing

 **CD 1** / Track 23

With its strong cries of *tregül, tregül, tregül* that provide its onomatopoeic *Mapudungun* name *tregül*, the Southern Lapwing characterizes diverse open habitats of South America. This regular-sized bird is common throughout South America, except in the dense forests of Amazonia, the high peaks of the Andes, and the dry deserts of northern Chile and Peru. In southern Chile it is common to observe the Southern Lapwing in the grasslands and wet fields adjacent to forests, from the Mediterranean region of the sclerophyllous forests in Central Chile to the sub-Antarctic region south of Tierra del Fuego, where it receives the Yahgan name of *pilit.* In the southern regions it lives together with flocks of Buff-Necked ibises and other birds while feeding from the soil on every type of invertebrate including larvae and adult insects.

The Southern Lapwing carries on its head a long and thin black crest. Its body stands out by being multi-colored: *i*) it is black in the front, the lower neck, chest and external tail-coverts; *ii*) light gray on the rest of its head and neck; *iii*) greenish-brown with a metallic shine on the upper parts of the wings; *iv*) white on the flanks and sub-terminal feathers; *v*) pink on the legs and beak; and *vi*) the iris of the eyes is an intense red. The Southern Lapwing displays all its coloration in its courtship flights and dances while it emits varied calls. This courtship display of the *tregül* inspires the dances of Mapuche ceremony of the *ngillatun**, which asks for good weather, harvests, and the well being of the people of the *mapu* (=land).

In the *Lafkenche* territory, the *tregül* are welcomed by the farming communities. The *tregül* are referred to as "meticulous cleaners," because they control the pests and populations of larvae. In addition, with its pecking the *tregül* "allows the soil to breath," contributing to the fertility of the land. Finally, their flights enable the farmers to forecast the weather: when the *tregül* flies low it is certain that it will rain.

The *tregül* is normally calm, but it becomes very aggressive when defending its nest or territory with its strong barbed wings. In the *Williche* territory, people call the *tregül* "vigilant bird" because, with its loud vocalizations, it announces the presence of any person or animal passing through the area. The *tregül* screams whenever it sees someone, even in the middle of the night. The *Lafkenche* and *Williche* people tell that this custom of the *tregül* makes them wonder if this bird ever sleeps. As the *Lafkenche* poet, Lorenzo Aillapan says:

119

WETLAND BIRDS

MAPUCHE STORY

CD 1/ Track 63

The Southern Lapwing is a sentinel, who upon observing raptors and other hunting animals, alerts the other bird species with its call *tregül, tregül, tregül....* Well armed with spurs on its wings, the Lapwing confronts the foreigners in its territory, but it also strongly expresses romance, its love of life, with a variation of its call: *triliu, triliu, triliu, triliu, trültriu, trültriu, trültriu, trültriu, trültriu.* With this call the male courts the female, dancing and singing. This courtship is represented in the Mapuche rite of *ngillatun*, where with open wings, like open arms, it dances counter-clockwise, calling out *trililililiuuuuuu* towards the north, *trililililiuuuuuu* towards the west, *trililililiuuuuuu* towards the south, *trililililiuuuuuu* towards the east. With this romantic song and its lover's dance the Southern Lapwing invokes the Universal Fertility Spirit.

MAPUDUNGUN VERSION

Tachi tregül (tero) lloftukellufe üñüm ka feyti kulliñ femngechi miyaukelu tralkatufe lelfün meu tie küpay pipingey tregül, tregül, tregül...... pu mürwen wera üñüm ellkautuy. Müna küme waykituley tañi epu müpü, tregül may kewa wemükefi kayñe akukelu ñi ad mapu lelfün meu. Welu kafeyti tregül duantulu pünowal müna küme ül nentuniekey fill wünül: triliu – triliu – triliu... trultriu, trultriu, trultriu... Femngechi feyti alka pünoy kude tregül pürüy ka ülkantuy. Katachi kawin dungu amuley feyñi chumken mapuche NGILLATUN, kefafay ka amuldunguy epu lipan nülay (müpü reke nülay) fey amuley rakiñ antü reke, wünül dunguley trililililiuuuu pikum püle trililililiuuuu lafken püle, trililililiuuuu – willi püle, trililililiuuuu tripan antü püle. Omfucha/ Omkude mütrüm adtuleyngün.

`The ngillatun is a special Mapuche ritual carried out to praise, ask or beg the four deities of the wenu mapu (land above) to maintain or restore the well-being of the people who inhabit the mapu (land) (see Catrileo 1998, p. 204).

Yahgan: Pílit

Mapudungun: **Tregül,** Queltehue,
            Queltreu, Treile

Spanish: **Queltehue, Treile, Tero,**
            Tero-tero, Fraile

English: Southern Lapwing

Scientific: *Vanellus chilensis*
            (Charadriidae)

Mapuche Territory

Yahgan Territory ⬜

Year-round Resident ⬤

Resident northern populations; extreme southern
populations migrate north during the winter
Common in open habitats
Sighting probability: 4

WETLAND BIRDS

IDENTIFICATION	12-14" (31-36 cm) Gray head and neck with characteristic black forehead, throat, and chest. Red eyes and legs. When flying it exhibits an unmistakable black and white design. The plumage of the upperparts is mainly brownish gray with a metallic bronzy sheen, especially notorious on the shoulders.
HABITAT	It inhabits a variety of open habitats, including grasslands, wetlands, farmlands, estuaries, river and lakes banks, as well as urban parks and gardens; from the sea level to 3,000 m in the Andean Cordillera.
HABITS	They are gregarious during the non-reproductive season. In contrast, during the reproductive season they are territorial, and live in pairs or groups of 3 to 5 individuals. Their alarm calls alert about the presence of predators, humans and dogs from a long distance. They nest on the ground, and the parents become very aggressive when a predator or another threat approaches the nest, and their 3 or 4 eggs, or chicks.
DIET	It eats a variety of larvae and adult insects, and other invertebrates.
CONSERVATION	Two subspecies inhabit the South American temperate forest biome: *Vanellus chilensis chilensis* (from Atacama to Chiloe in Chile, and from Jujuy to Neuquen in Argentina), and *V.c. fretensis* from Aysen (Chile) and Santa Cruz (Argentina) to Navarino Island and other islands south of Tierra del Fuego. The regional distribution and the abundance of both subspecies have increased due to deforestation associated to the expansion of cattle ranching and agricultural lands, since neither of these subspecies inhabit dense forest habitats. The species is classified as *Least Concern* by IUCN.

BIRDS OF THE
FOREST MARGINS

Kono
Torcaza
Chilean Pigeon

 **CD 1** / Track 24

In the forests of southern Chile and Argentina, inhabits a Chilean Pigeon or *Kono*, an endemic pigeon larger than the domestic one so common in the world's cities. *Kono* has a beautiful, reddish-chestnut coloration, orange eyes, and an elegant white band at the nape of the neck with a metallic green patch below. It is gregarious, and lives in flocks high in the trees where they eat fleshy fruits like the peumo (*Cryptocaria alba*), the lingue (*Persea lingue*), the Winter's Bark (*Drimys winteri*) or the olivillo (*Aextoxicon punctatum*). They nest in the trees, constructing small platforms of small, dry sticks, where they incubate and then feed their chicks with a kind of "milk" from the digested seeds of fruit.

Hidden in the foliage of the trees, the pigeons emit their sonorous cooing that so typifies the austral forests—the sound was heard by Spanish conquistadors and caused them to believe that *kono* was the most abundant bird. Places such as Conumo (37°16'S; 73°14'W) in the mountains near Arauco, and the town of Pucón (39°15'S; 71°58'W) on the shores of Villarrica Lake express with their names of *Mapudungun* origin, the abundance of the Chilean Pigeon. Conumo means "with (= *meu*) pigeons (= *kono* or *konun*)", and Pucón means many Chilean pigeons (*pu* = plural prefix; *kon* = abbreviation *kono* or *konun*). As described by the poet Lorenzo Aillapan, the Chilean Pigeon or *kono* was so abundant in past times that its flocks covered the sky with winged clouds—clouds that crossed through the *ankawenu'* or aerial space of the Mapuche territory.

During the first decades of the 20th century, the Chilean Pigeon became the favorite prey of modern hunters in southern and central Chile. They waited to hunt it in the month of April, when the lingue (*Persea lingue*) bears fruits. When pigeons gorged on the lingue or *linge* fruit, its flavor permeated the flesh of the Chilean Pigeon or *kono*, which acquired a flavor highly-prized by the modern hunters and their customers. The over-hunting that ensued to satisfy the appetite for this flavor, together with habitat degradation and the devastating effect of the Newcastle epidemic virus, carried the Chilean Pigeon to the edge of extinction. Fortunately, there is still hope: *kiñekelewey kono mawida mew*, there are still a few wild pigeons flying in the austral forests.

Today its hunting is forbidden and the populations of Chilean Pigeon have begun to recover. It is possible to observe again flocks of more than 30 individuals. Even more important, the existence of

BIRDS OF THE FOREST MARGINS

kono and its habitat are beginning to be respected again, and the diverse, human cultural viewpoints in southern Chile and Argentina are being revalued today.

'The *ankawenu* is one of the four levels in the organization of space according to the Mapuche worldview. It corresponds to the aerial space between the *wenumapu* or celestial space, and the *mapu* or the land where we live.

Yahgan: out of bird's range

Mapudungun: **Kono,** Konun, Turcasa, Cono

Spanish: **Torcaza**, Paloma araucana

English: Chilean Pigeon

Scientific: *Patagioenas araucana*
                  (Columbidae)

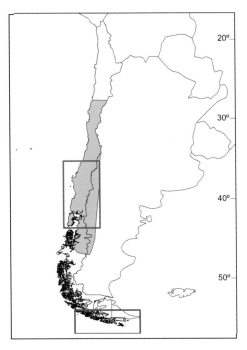

Mapuche Territory ☐

Yahgan Territory ☐

Year-round Resident ⬤

Resident northern populations; southern populations
migrate north during the winter
In a variety of forest types
Sighting probability: 3

IDENTIFICATION	13-15" (34-38 cm) The largest native pigeon. On its nape, it has a characteristic white, narrow semicollar, with metallic green plumage below. General purple brown plumage, except for the upperparts' gray plumage on the back and coverts on the wings. Striking red iris and legs.
HABITAT	It inhabits the sclerophyllous woodlands, Valdivian rainforests, North Patagonian and other forest types, as well as the vicinity of farmlands, between Vallenar (28°S) and Taitao (47°S).
HABITS	Gregarious, and its flocks feed and nest on the canopy of trees. Their nests consist of a platform of sticks, placed among the branches of a medium-height tree, where they lay one or two eggs per reproductive season. Its cooing is a characteristic deep doubled *hooo-HOOOO hoo-HOOO*, preceded by a guttural sound PRRRRR.
DIET	It feeds on a variety of fruits on trees, such as lingue (*Persea lingue*), boldo (*Peumus boldus*), peumo (*Cryptocaria alba*) and mañío (*Podocarpus nubigena*), or shrubs, such as maqui (*Aristotelia chilensis*) and quilo (*Muehlenbeckia hastulata*).
CONSERVATION	Vulnerable species (according to the Chilean Red Book of Fauna), and endemic to the temperate forests of South America (its distribution is restricted to the Endemic Bird Areas 060 and 061, Birdlife International). ⓥ

Maykoño
Tórtola
Eared Dove

 **CD 1** / Track 25

According to the Mapuche ornithological view, this small gray dove with black spots and bands on the wings is one of the half forest (*mawida*), and half prairie (*lelfün*), birds. Every sunrise, flocks of Eared Doves or *maykoño* fly from the woodlands to the prairies, feeding places where they search for any type of seed and for ponds to drink water. Each sunset the flocks return to their trees in the woodlands, where they have their nesting and roosting site. This daily cycle creates Eared Dove flyways that the hunters know very well to wait with their rifles for the flocks that pass flying in the early morning and late afternoon of each day. Fortunately, the Eared Dove is very prolific with up to three clutches per year, and their populations remain abundant in Chile and Argentina, as well as almost all of South America (except the Amazon Basin and the Andes above 3,000 m). It even has become a frequent inhabitant of villages and urban centers. The Eared Dove is today a native pigeon in the central plazas of cities like Temuco or Chillán in the south of Chile, or in Quito or Bogotá in northern South America.

The Eared Dove, as with the domestic pigeon, dedicates much of its time to reproduction. It constructs its nests of little sticks and feathers, which often fall onto the soil. This circumstance reflected in the *Williche* name of Muicolpue, in the Coastal Mountain Range near Osorno, which indicates place (*ue*) of dove (*muiko*; abbreviation of *muikoño*) feathers (*lp*; abbreviation of *lepi*). As such, *maykoño* accompany the rural and urban human communities, and from their nests and roosting sites comes their characteristic song, the typical cooing of the doves. Its call and the long period it spends on its nests are poetically described and interpreted by the *Lafkenche* Bird Man, Lorenzo Aillapan.

127

BIRDS OF THE FOREST MARGINS

MAPUCHE STORY

 CD 1/ Track 60

Maykoño is a feminine name, and with its pitiful call (*hooo-hooo-hooo...*), the Eared Dove has cried since humanity has existed. It is a spokesperson for the pain of existence, suffering for the frequent fall of its little eggs or chicks from the rudimentary nests, constructed with a few little branches. It cries with the pain of other birds, other living beings and of the Earth itself, hurt by human actions, such as the extensive clear-cutting of forests and the excavations of mines. It cries for natural furies, like volcanic eruptions, earthquakes and tidal waves. The dove or *maykoño* reminds us with its cooing (*hooo-hooo-hooo...*), that pain is as integral a part of life as is happiness and pleasure.

MAPUDUNGUN VERSION

Tachi maykoño domo üyngey ka niey ngüman ülkantu: uy-uy-uy-uy kogon-kogon kogon-kogon uy-uy-uy-uy. Trangoy-kuram-püñeñ-kogon. Tachi maykoño ngümakey deuma mülen mogen che. Ülkantun meu nentuniekey mongen duam, trangoy kuram, trananagita püñeñ rume pichin kochay niey ñi dañe, katachi kutrantulelu ka wera üñün, kulliñem mongelelu katachi ñukemapu allfülngelu fillfemün pu wedake che, katrün mawidantu ka kintugen fill duamfal pañillwe, ka fey illkun ad mapu, kütral degiñ, fütranüyün ka tripako lafken. Fey ngümay weñagütu tukulpay küme ül uy-uy-uy-uy.... kutrantunlladkün weraley mongen fey ka wefrumetuy ayiwün ka llakong duam.

Yahgan: out of bird's range

Mapudungun: **Maykoño,** Cullpo, Culpo, Mayconu, Muikoño, Cono

Spanish: **Tórtola**, Torcaza

English: Eared Dove, Spotted Dove

Scientific: *Zenaida auriculata* (Columbidae)

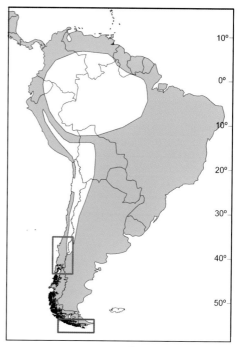

Mapuche Territory ☐

Yahgan Territory ☐

Year-round Resident ⬤

Common in forest edges and open habitats, including farmlands
Sighting probability: 4

BIRDS OF THE FOREST MARGINS

IDENTIFICATION	9-11" (23-28 cm) Gray, purplish brown plumage, with black spots on the head and wings. Sides of neck with iridescent, pink or golden-green feathers. Black beak and pink legs. Male has a paler grayish crown.
HABITAT	Inhabits forest edges and open forests, as well as wide variety of open habitats, including farmlands, urban parks and gardens, in humid, semiarid, and arid regions, on the coast, central valley, and Andean cordillera, up to 2,500 m. It is very common from Coquimbo to Aysen, and occasional visitor in Tarapaca in the north, and in Magallanes in the south, where it is a summer visitor in Tierra del Fuego.
HABITS	Gregarious, during the non-reproductive season forms large flocks which feed on the ground in farmlands and other open habitats. The Eared Dove perches in trees where it builds its rudimentary nests with sticks. It emits a melancholic, soft, high-pitched cooing, *hooo-hooo-hooo*.
DIET	Feeds on a variety of seeds taken from the ground.
CONSERVATION	Eleven subspecies have been identified for *Zenaida auriculata*. The South American temperate forests are inhabited by the subspecies *Z.a. auriculata*, which in spite of the hunting pressures does not present conservation problems, and is classified as "Least Concern" (IUCN).

Pütriu
Pitío
Chilean Flicker

CD 1 / Track 26

The Chilean Flicker is the most common of the Chilean woodpeckers, and typifies the diverse forest and shrubland habitats with its resounding territorial call *pitrío, pitrío, pitrío* ..., which gives origin to the onomatopoeic *Mapudungun* name *pütriu*.

The *pütriu* occupies a great variety of habitat types—in fields with few trees, in regenerating forests, and in old-growth forests—from sea-level to 2,000 m elevation. In the forests the flickers also devour berries and seeds, especially from the *quilas* or bamboos *(Chusquea* sp.). The *Lafkenche* say that the *pütriu* also feeds on sap "the life drawn from the mountains through the trees *(feyti mongen aliwen mauwida meu)*."

In contrast with the other austral woodpeckers, the *pütriu* feeds not only on trunks, but also in the soil where it captures a great variety of worms, larvae, eggs, nymphs, and adult insects, including ants. It also not only nests in tree cavities, but also in cavities it excavates on cliffs. Hence, like the Eared Dove or *maykoño*, the *pütriu* is a half-tree/half-soil bird.

It is abundant in the fields of the *Lafkenche* territory, and the people point out that the *pütriu* plays an important role in the community of birds, because with its loud alarm call it announces danger. As the Mapuche Bird-Man, Lorenzo Aillapan says:

131

MAPUCHE STORY

 CD 1/ Track 58

The *pütriu* notifies the other birds with its shout of alarm *trü trü trü trü trü* the presence of the hunter. With its other shout *trü tü tü tü tü* it notifies that the hunter has gone and that the half tree-half ground birds can return with safety to the prairie. As such, the communication of this carpenter bird expresses the spirit of solidarity that exists between communities of birds and which beautifies Mother Nature.

MAPUDUNGUN VERSION

Tachi pütriu inkaniekey pu mürwen üñüm wakeñ ül men pütriu-pütriu-pütriu-pütriu-uuuu... tie küpaytralkatufe, feytañi wakeñ ül meu: pütriu-pütriu-pütriu-pütriu-uuuu tie amutuy tralkatufe feula wiñotuay pu wera üñüm rangiñ mawidantu ka rangiñ lelfün küme pitrongtuay wera kachu lelfün meu. Feymeu tachi katafealiwen üñüm kiñewayiñ dungu mingako püllü weftuay taiñ üñüm kiñewayiñ dungu mingako püllü weftuay taiñ üñümngen ka fey meu rume adtuay taiñ ñuke mapu.

Yahgan: out of bird's range

Mapudungun: **Pütriu,** Pitio, Pitihue, Pitiu, Pitigüe, Rere, Pütiw

Spanish: **Pitío**, Pitigüe

English: Chilean Flicker

Scientific: *Colaptes pitius* (Picidae)

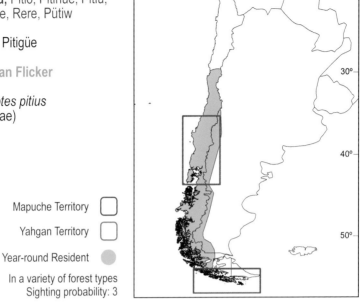

Mapuche Territory

Yahgan Territory

Year-round Resident

In a variety of forest types
Sighting probability: 3

IDENTIFICATION	11-14" (29-35 cm) Forehead, crown, and nape are gray. Sides of head and throat are beige. Iris is yellow in adults, and light blue in immature birds. Plumage on upperparts are brown, grayish barred with white. Distinctive white rump, seen more clearly when it flies. Pale beige belly.
HABITAT	It inhabits open forested areas, forest margins, and shrublands from the coast to the pre-Andean cordillera (2,000 m), from Neuquen (38°S, Argentina) and Huaco Valley (28°S, Chile) to the Magellan Strait (53°S).
HABITS	Often seen in pairs or in family groups. It has a characteristic undulating flight, alternating five flaps and then glides with wings closed. In contrast with other woodpeckers, it not only nests in tree cavities but also on cliffs. It feeds on insects, not only on the bark of trees, but also on the ground, in grasslands and steppes. It repeats a loud characteristic call *"pit-tweeo."*
DIET	Feeds on adult and larval insects.
CONSERVATION	Distribution restricted to the Endemic Bird Areas 060 and 061 (Birdlife International), but is classified as a species of Least Concern (IUCN).

BIRDS OF THE FOREST MARGINS

Pichi rere
Carpinterito
Striped Woodpecker

(•) **CD 1** / Track 27

The ancient Mapuche called the family of woodpeckers *katanaliwenüñüm*, alluding to the birds' (*üñüm*) habit of perforating (*katan*) the trees (*aliwen*). The Striped Woodpeckers are the smallest representatives of this group of birds, from which comes its *Mapudungun* name *pichirere* (*pichi* = small; *rere* = woodpecker). Its scientific name, *Picoides lignarius*, means "similar to the wood cutter" and this little woodpecker, indeed, is recognized by its rapid and constant pecking. In the countryside, people know it as "picamaderos" or "wood chopper."

The male has a characteristic red forehead and crown. He marks his territory with a rapid succession of pecks against a dry trunk, producing the sound that ornithologist Guillermo Egli describes as "drumming." To build their nests, the Striped Woodpecker excavate small cavities in the trunks of old, living or dead standing trees. In many types of forest, the *pichirere* laboriously moves up and down tree trunks, supported by their tail, extracting larval and adult insects with its strong beak. The Mapuche Bird-Man tells the story of this member of the *katanaliwenüñüm* family:

MAPUCHE STORY

 CD 1/ Track 59

Striped woodpeckers or the *pichirere*, the same as the *pütriu* and the *rere*, pass the day pecking on the trunks of trees, extracting the larvae, which likewise feed on the sap and the wood of the plants. As such, the flow of life is established between the sap of the plants, the larvae of the insects and the woodpeckers, who resound in the forests, or *mawida*, of the Mapuche lands.

MAPUDUNGUN VERSION

Tachi pichi katafealiwenüñum pichirere, ka feyti pütriu ka ti rere, rumel pünalekey pitrong-pitrong aliwen mamüll, fey nentual piru merun fid mongepiyüm fill aliwen ka añumka. Femngechi, epuñpüle mongeluwi tuwikülenmeu fill anümka ka feyti wera piru fill idike fey wepumniey tachi pu katafealiwenüñum fey negümnegümngey mawida meu cheuñimülen pumapuche.

Yahgan: out of bird's range

Mapudungun: **Pichi rere**

Spanish: **Carpinterito**,
                Carpintero bataraz grande

English: Striped Woodpecker

Scientific: *Picoides lignarius*
                (Picidae)

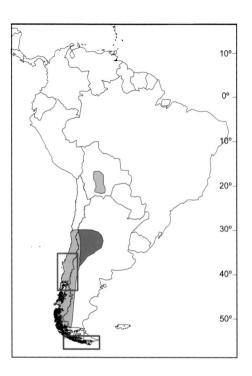

Mapuche Territory ☐

Yahgan Territory ☐

Year-round Resident ●

Occasional Visitor ●

Common in sclerophyllous, Valdivian,
and other types of forests
Sighting probability: 3

IDENTIFICATION	6-7" (15-19 cm) Upperparts are transversally barred with white and black. Underparts, whitish yellow with fine, longitudinal black barring. Male has a red nape, black crown and a central red spot. On the sides of its head, both sexes have a distinct black line that runs through the eye.
HABITAT	Diverse types of forested habitats, orchards, and shrublands, as well as urban parks. Found from Neuquen (36°S, Argentina) and Coquimbo (29°S, Chile) to the area of Puerto Natales (52°S) in the Magallenic region, from the sea level to the Andean pre-cordillera (2,000 m).
HABITS	Usually in pairs. Short undulating flights from tree to tree, moving nimbly between branches and especially dry trunks searching for small insects. Nests in tree cavities. It emits a characteristic low and fast drumming, when it pecks on trees.
DIET	Feeds mainly on adult and larval insects.
CONSERVATION	Distribution restricted to the Endemic Bird Areas 060 and 061 (Birdlife International), but classified as a species of Least Concern (IUCN).

Feo
Wiyu
Fío-fío
White-Crested Elaenia

 **CD 1** / Track 28

With its crest of white plumage, and double-whistle call during the breeding season, the White-Crested Elaenia typifies the austral forests of southern Chile. The double whistle of this little, olive-green bird is imitated by the *Mapudungun* name *fiu-fiu* which was translated as *fío-fío* by the Spanish conquerors. The call also inspired the name of the largest river in the South American temperate forest region, the Bío-Bío River. The single whistle of this bird is often heard in the early morning and gives rise to its other *Mapudungun* name: *wiyu.*

White-Crested Elaenia stands out among austral avifauna for its remarkable migratory behavior. It spends the winter in tropical cloud forests of Ecuador, Peru and Brazil. At the beginning of spring, adults begin their flight south and go as far south as Cape Horn, where they nest and feed their young, which then fly with them to the tropics at the end of summer. The migratory flight of the White-Crested Elaenia is so precise that, even having flown 5,000 kilometers, they arrive to southern Chile and Argentina and nest in the same trees year after year. It often builds its nests and feeds on the fruits of a tree that is sacred to the Mapuche people, the *foye* or Winter's Bark (*Drimys winteri*).

The White-Crested Elaenia has a generalist diet. Throughout the year, it is a skillful insect hunter, but during the flowering period of the *notrü* or firebush (*Embothrium coccineum*) it acts like a hummingbird, flying from flower to flower, drinking nectar. When the plants begin to fruit in summer, it eats the berries of many trees and brushes, such as Box-Leafed Barberry *(Berberis buxifolia)* and Wild Currant *(Ribes magellanicum).*

In the forests and mountains of the Mapuche territory, where berries abound, the whistles of the wiyu also inspire many entertaining stories that are told between women. Poet Lorenzo Aillapan tells us that:

137

MAPUCHE STORY

 CD 1/ Track 57

Near the coast of Temuco, the *Lafkenche* associate the whistle of the White-Crested Elaenia, the *wiyu,* with the whistles that young men direct towards women when they go to collect *maki* (*Aristotelia chilensis*) and other fruits in the forest. It is told that some women have recovered from sickness when they receive the love of these young men. Humorously, a beautiful grandmother from Cunco, towards the mountains from Temuco, went one morning to collect water at a well where she heard these flirty whistles *feeeo, feeeo, feeeo...* Amazed she said, "it seems that I am still attractive." She emptied the bucket of water and returned to the well right away, where she only found not a young lover but the roguish *wiyu* instead, whistling its characteristic call.

MAPUDUNGUN VERSION

Pu lafkenche, konün antü püle Temüko waria, keñangekey wiykeñ wiyu pengeyüm fütra pura domo miyaukelu domokulme wechewentru (maqui) makemeyalu fill pu domo mülekelu kam fill fün mawida. Fey piam kiñe ülchadomo llaftuy fechiwüla wentrutulu, rumel tremotuy - fey may pi: wentru mayta lawenwürke. Allangechi, kiñe kuku mülelu kunko (Cunco) waria püle, pirentu winkul Temüko püle, kiñe liwen amuy komealu kotuwe meu fey allkütuy wiykeñ wentru domokulngen. Feynga konduamlu feypi "Petu may illufalngeperkelan küme trawa kam nielu inche," witruy ñi medenko fey ka wuñotug tañi kotuwe meu, welu feynga rewedwed wiyu üñüm fey wiykeñ wiykeñ ülkantu mekerkey.

BIRDS OF THE FOREST MARGINS

Yahgan:	**Feo**, Fuiiū, Puiū, Pouyou
Mapudungun:	**Wiyu,** Fio, Fiu-Fiu, Fio-fio, Vio-vio, Peutren
Spanish:	**Fío-fío,** Fío-fío silbón
English:	White-Crested Elaenia
Scientific:	*Elaenia albiceps* (Tyrannidae)

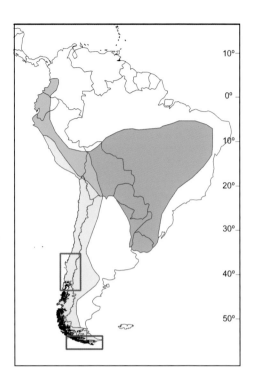

Mapuche Territory ☐
Yahgan Territory ☐
Year-round Resident ●
Summer Range ●
Winter Range ●

Migrate north to Amazonian forests during the winter
In a variety of forest types, during spring and summer
Sighting probability: 4

IDENTIFICATION	5-6" (13-15 cm) Distinct white, erect crest on the crown. Upperparts have a gray-olive color, and its dark wings have two white stripes, and blackish tail. Belly is whitish with yellowish touches.
HABITAT	Live in open and dense forests, as well as shrublands, parks and gardens. Found from the sea level to the Andean pre-cordillera (2,000 m).
HABITS	Alone, in pairs or in mixed flocks with other passerines. Emits a characteristic whistle in the spring-summer austral forests, "*feeeo-feeeo*". The subspecies *Elaenia albiceps chilensis* inhabits from Copiapo (27°S) to Cape Horn and Diego Ramirez Islands (56°S). This subspecies migrates south in the spring, which begins in September-October for most of Southern Argentina, and central and southern Chile. At the end of March, the beginning of fall, it migrates to the Amazonian forests of Brazil and Peru, as well as other forests of tropical South America.
DIET	It feeds on flying insects, small fruits and the nectar of flowers. Sometimes resembles a hummingbird sipping nectar, however the White-Crested Elaenia pierces holes in the base of the visited flowers.
CONSERVATION	It includes six subspecies, but the only one present in the South American temperate biome is *E. albiceps chilensis*. It is classified as a species of Least Concern (IUCN).

Twín
Chidüf
Jilguero
Black-Chinned Siskin

CD 1 / Track 29

The Black-Chinned Siskin stands out due to the prolonged call and showy colors of the male. The male's black crown and throat are denoted in its scientific name *Carduelis* (siskin) *barbata* (bearded), and in its Argentine name "little black head." Before and after the nesting season, they form large flocks that bring music to the austral forests as far south as Cape Horn. In the Yahgan territory, stories of the Black-Chinned Siskin, or *twin*, relate the beauty of their song, behavior, and color of these little birds to their descent from young maidens.

141

BIRDS OF THE FOREST MARGINS

YAHGAN STORY

 CD 2/ Track 5

When we were children, they told us we should not cherish any piece of wood, stone or anything like that, because it could turn into a person. To this girl, it happened. She picked up a beautiful stone she had found in *Tashuani* on Hoste Island at the shore of the sea. And she said to it, "This is my baby." She cared for it until it transformed into her baby. A little head grew, as well as hands and human feet; his body was made of stone. She had it always with her, took it to her breast, until it bit a chunk from her. So the father of the girl said to her, "What we shall do is throw it into the water because it is bad. We cannot take it with us." So they threw it into the sea.

They thought themselves lucky that he had died already, and they came to a place where they set up a hut, and at night he arrived again, always following his mother. Then, the little stone grew up, and when a man, he caught the women who crossed the canal in canoe, and he killed the men that opposed him. The others ran away. He had many women; it was here at *Lum*˙. Up in the mountains of *Umushupuliak*, the stone giant tore out big *katran*˙˙ trees with their roots, carried them on his shoulder for the women to eat.

One day on the mountain, he got a thorn in his foot, and later he got ill. He said, "Oh, my foot is hurting. Take out the thorn!" He lay down in the meadow, and the women pierced into the spot with their *ami*˙˙˙ where the thorn was. They hurt him, and he cried out, but it was such a nice, sunny day that he slept and kept on sleeping. The women decided to burn him. With the same trees that he had brought, they surrounded him. The branches were dry, and the fire blazed. His head, hands and feet burned, and his body, which was of rock, began to perish. Small stones jumped out of the fire, and the women put them back into their place with sticks. The women that had touched these stones turned into birds, into Black-Chinned Siskins, or *twin*. Siskins are green and yellow in color, and you can see them in September. When all was burned, they flew away singing their song, joyful to have been liberated from the giant.

This rock that was the giant is still in *Lum*, and when you make a fire near to that place, little stones begin to explode all around.

˙*Lum is a place on the northern coast of Navarino Island at the Beagle Channel.*

˙˙*Edible fungus from Cyttaria genus.*

˙˙˙Tool made from a tibia (a leg bone) of a marine bird (e.g., cormorants, seagulls) by Yahgan women to weave baskets made of rushes.

Yahgan: **Twín,** Tuwín, Dwīin, Duīin

Mapudungun: **Chidüf,** Quechan, Chirihüe, Dihue

Spanish: **Jilguero,** Cabecitanegra austral

English: Black-Chinned Siskin

Scientific: *Carduelis barbata* (Fringillidae)

Mapuche Territory ☐

Yahgan Territory ☐

Year-round Resident ◯

Occasional Visitor ●

In a variety of forests types and open habitats
Sighting probability: 4

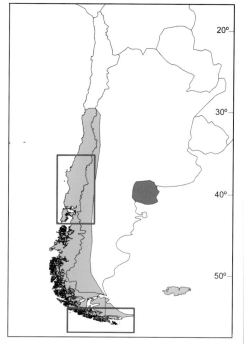

IDENTIFICATION	4.7-6" (12-15 cm) Green-yellow colored bird. Sexual dimorphism: The male has a distinctive, black crown and throat. The female is paler, less bright and has a gray crown.
HABITAT	Found in the forest interior and margins of dense and open forests, as well as shrublands, farmlands, urban parks and gardens. From the Huasco Valley (28°S) throughout the South American temperate forest biome down to the Cape Horn Archipelago (56°S); from the sea level to the Andean Cordillera (3,000 m).
HABITS	In pairs or small groups during reproductive season in spring, while it's found in large flocks during winter. Granivorous and insectivorous, it feeds on the ground or in the canopy. Its song is quick with many notes doubled or repeated: *"chwee."*
DIET	It feeds mainly on seeds, and fruits.
CONSERVATION	Its distribution is restricted to the Endemic Bird Areas 060 and 061 (Birdlife International), but is classified as a species of Least Concern (IUCN).

143

Tashúr
Kichan
Cometocino
Patagonian Sierra Finch

 CD 1 / Track 30

With its eye catching yellow breast and blue-grayish head and wings, the Patagonian Sierra Finch inhabits the whole range of *Nothofagus* forests in southern Chile and Argentina. This colorful bird usually perches in the tree canopy, and one often hears the territorial song of a solitary male coming from high in the branches. At other times, it is possible to distinguish the dry, "*chic-chic-chic,*" of flocks keeping track of one another as they disperse among the foliage of trees or bushes. It inhabits the interior and margin of the forests, where it feeds on a great variety of seeds, fruits, insects, and even robs nectar by pecking a little hole at the base of flowers, such as fuchsia or *chillko* (*Fuchsia magellanica*), firebush or *notrü* (*Embothrium coccineum*) and the vine *kolkopiw* (*Philesia magellanica*).

On Chiloé Island, the Patagonian Sierra Finch, or *kichan*, is considered a messenger who brings good news with its call , "*chic-chic-chic,*" like the striking of two stones together. It brings bad news when it comes near the houses and sings its penetrating and insistent whistle, "*chuif-chío, chuif-chío, chuif-chío.*"

One of the most abundant permanent resident birds of the temperate rainforests between Chiloe Island and Cape Horn, the Patagonian Sierra Finch, gives color and life to the months of snow. Perhaps for this reason, the Yahgan grandmothers Úrsula and Cristina Calderón consider the Patagonian Sierra Finch, or *tashúr*, one of the most pleasing birds.

BIRDS OF THE FOREST MARGINS

Yahgan: **Tashúr**, Tachurj, Tõšur˜

Mapudungun: *Kichan*, Chuchan

Spanish: **Cometocino**,
Comesebo patagónico

English: Patagonian Sierra Finch,
Sierra Finch

Scientific: *Phrygilus patagonicus*
(Fringillidae)

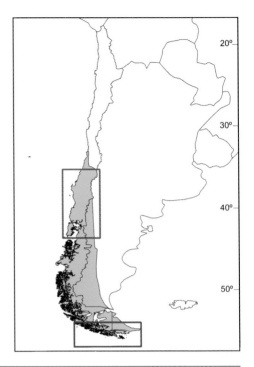

Mapuche Territory ☐

Yahgan Territory ☐

Year-round Resident ⬤

In a variety of forest types
Sighting probability: 4

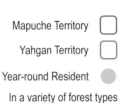

IDENTIFICATION	5.5-6.2" (14-16 cm) Head, neck, wings, and tail are gray-bluish. Sexual dimorphism: The male has a bright, yellowish brown back, and a bright yellow belly. The female is more dull with less distinct coloration.
HABITAT	Forest interior and margins of dense and open forests, as well as shrublands, steppe, and urban gardens. From La Serena (30°S) throughout the South American temperate forest biome down to the Cape Horn Archipelago (56oS); from the sea level to the Andean pre-cordillera (1,800m).
HABITS	In pairs or flocks. Active singer during its breeding season, its musical and repetitive song, *"fweep-fweeeo, fweep-fweeeo, fweep-fweeeo,"* is emitted when it perches at the top of trees. Its alarm call, *"chick-chick-chick,"* that emerges from the middle canopies and shrubs is also frequently heard inside the forests. It nests in dense thickets.
DIET	Feeds mainly on seeds, and fruits, and occasionaly on nectar. It is considered a nectar robber because, to get the nectar, it pierces holes in the base of the visited flowers; hence, it does not pollinate the visited flowers.
CONSERVATION	Distribution restricted to the Endemic Bird Areas 060 and 061 (Birdlife International), but is classified as a species of Least Concern (IUCN).

Hakasir
Wilki
Zorzal
Austral Thrush

 **CD 1** / Track 31

The melodic call of the Austral Thrush epitomizes the sunrises and sunsets of the forests, farmlands, and even plazas and gardens of southern South America. From its song derives the *Mapudungun* name *wilki*, as well as its nickname "orchestra director of the birds and nature," given to it by the *Lafkenche* poet Lorenzo Aillapan. In San Juan de Chadmo on Chiloe Island, its song is interpreted, "blessed and praised be God." For the *Lafkenche* poet the melodious *wilki* captures the essence of the Earth, revitalizing all the living beings who listen to its whistle:

Singing I say: life-happiness-love. Among songs and dances, through the air to the ear, I say hello to the human, likewise to those who care for the children, to those who are living thanks to virgin Nature. *Chülle Mapu*, earthly paradise, from the mountains to the sea From north to south, all things rejuvenate, returning to spirituality...*	Pin ñi ülkantun: mongen, ayüwün, poyen. Rangiñ ülkantu ka pürün kürüf mew wefi pilun püle, Chaliwün che feyti petu penielu fey ñi pichike pu yall, Feyti petu mongelelu puche, we mongen lelfün füla. Chülle mapu dulliñ mülewe rumel pirentu winkul ka lafken püle Pikum püle ka willi kürüf rüf mongetuy wüñotulu püllüyem…

The *wilki* is the largest flying Passeriform in the *Nothofagus* forests, where it occupies all habitat types in the interior and edges of forests. They consume all kinds of food from the soil and forest canopy. The *wilki* is so abundant throughout the Mapuche territory, that it gives its names to places in the northern, central, and southern regions. For example, in the Cachapoal Valley, Huilquío, means place of Austral Thrushes, (*wilki* = Austral Thrush, and *o* = diminutive of *we* or place), in the Andes Cordillera of Talca, Huilquilemu, or *Wilkilemu*, means forest (*lemu*) of the *wilki*, and on Chiloe Island, Huilqueco means river or water (*ko*) of the *wilki*.

In the Yahgan territory, the Austral Thrush, or *hakasir,* also inspires the women and other members of the culture. The Yahgans observe how this bird feeds and educates its chicks in trees and along the rivers that run through the forests. The Yahgan grandmothers Ursula and Cristina Calderón explain how, when the Austral Thrush has chicks, the mother teaches the oldest one. Grandmother Ursula describes how the "mother *hakasir* climbs to a stick. I saw her talking, whistling, teaching her child, constantly moving around the canopy. I was lying down, watching while the *hakasir* told him, 'when you get a brother, you'll have to educate him, and stimulate him to work and wash himself.' In that way the *hakasir* grows. If the Austral Thrush's son doesn't listen to his mother, if the little *hakasir* doesn't obey, she abandons him. The young one stays alone, and doesn't have anything to eat. Then, it feeds on bad *katran* (the fungus *Cyttaria sp.*), those that are overripe, that do harm, and the little *hakasir* dies. If the little *hakasir* obeys his mother, though, he grows up healthy eating good fruit like *amai* (*Gaultheria mucronata*) and *mema* (another fungus species of the genus *Cyttaria*)."

Thus, the Austral Thrush, *hákasir* or *wilki,* expresses to us how the lives and languages of the birds are interwoven with those of human beings in the forested Yahgan and Mapuche territories of southern South America.

YAHGAN STORY

 CD 2/ Track 11

When the Austral Thrush has chicks, the mother teaches the oldest one. She rises to a stick. I saw the little bird, teaching, talking to her child. I was lying down, watching her. She told him, "When you'll have a brother, you'll have to help raise him, stimulate him to work, to wash himself. In that way the little bird was singing. They flew to the river, where they dove into the water, washing themselves, shaking themselves, and flying back to the tree. In that way they grew up."

If the little Austral Thrush doesn't listen to his mother, if he doesn't obey, she abandons him. The young one stays alone, doesn't have anything to eat, lives on bad *dihueñes*, those that are already overripe, that do harm, and he dies. But on the other hand, if he obeys his mother, he grows up healthy living on good fruit, like *amai* (*Gaultheria mucronata*) and *mema* (another fungus species of the genus *Cyttaria*).

MAPUCHE STORY

With its call the austral thrush typifies the sunrises of the forests throughout the south of Chile and Argentina. The Mapudungun name *wilki* is onomatopoeic from its melodic whistle. To its beat grow the children, and the mothers nurse the babies singing, "*moyoy ta puñeñ, with! lüchütuy püñeñ, will!*"

Mapudungun version

Tachi wilki pu aliwen koniñi wakeñ ülkantun mawidaantu meu willi püle Chile mapa ka Argentina. Uytukun wilki fey tañi wünül ülkantun wüykeñkantu wünül tremün pichike domo ka feyti pu ñuke moyolu tañi pu püñeñ ülkantuy moyoy ta püñeñ – wüy/wüy ta püñeñ – lüchituy ta püñeñ.

Yahgan: **Hakasir**, Hakásij, Hakásirj, Xakacir

Mapudungun: **Wilki**, Huilqui, Huilque

Spanish: **Zorzal**, Zorzal patagónico

English: Austral Thrush

Scientific: *Turdus falcklandii* (Turdidae)

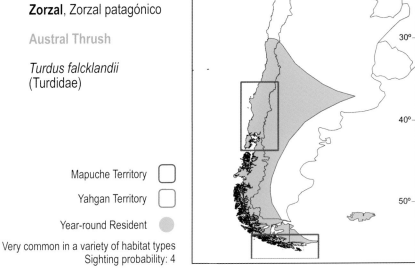

Mapuche Territory ☐

Yahgan Territory ☐

Year-round Resident ●

Very common in a variety of habitat types
Sighting probability: 4

BIRDS OF THE FOREST MARGINS

IDENTIFICATION	9-11" (23-29 cm) Dark brown head, back and wings while its underside is lighter. The throat is streaked with white with black. Juveniles have prominent black mottled underparts. Noticeable yellow bill, legs, and eye ring.
HABITAT	Found in the forest interior, and margins of dense and open forests, as well as shrublands, farmlands, urban parks and gardens. From Chañaral (27°S) throughout the South American temperate forest biome down to the Cape Horn Archipelago (56°S); from the sea level to the Andean Cordillera (2,500m).
HABITS	Alone or in pairs during spring and summer, and frequently observed in flocks during the non-breeding season. It feeds on the ground of open and closed habitats, as well as in the foliage; with its melodic whistle characterizes the austral sunrises and sunsets.
DIET	It feeds on earthworms, insects, and other invertebrates, as well as fruits.
CONSERVATION	Distribution restricted to the Endemic Bird Areas 060 and 061 (Birdlife International), but is classified as a species of Least Concern (IUCN).

BIRDS OF THE FOREST MARGINS

Chílij
Chedkan
Chercán
House Wren

(•) **CD 1** / Track 32

The House Wren has a very wide distribution, stretching from the forests of Canada to those of Cape Horn, although this range could include several subspecies or even cryptic species. This little Passerine occupies a great variety of habitat types. In the forests, it nests in trunk cavities. Although the holes have very narrow openings, inside they resemble true caverns. From this nesting behavior derives its scientific name *Troglodytes* (troglodyte, someone living in caverns) *aedon* (little). This behavior also might be associated to the Yahgan view of the House Wren, or *chílij*.

In ancestral times when birds were still humans, *chilij* possessed numerous wives (*kipa* = woman) that were representative of several bird species that also nest in cavities, like *lasijkipa* (the Swallow-woman) or *tatahurjkipa* (the Tree-Runner-woman). It is noticeable how *chilij* is constantly hunting insects to feed its chicks during the reproductive season. Chilij has as many as three clutches per year.

The House Wren is also characterized by its melodic vocalizations that include the trilled song of the male during the reproductive season, as well as numerous raspy vocalizations, such as the territorial call from which its *Mapudungun* name, *chedkan*, derives. Hence, as in the case of many other forest birds, the *Mapudungun* name is onomatopoeic. In turn, the Spanish name, *chercán*, derives from *Mapudungun*. In this manner, the onomatopoeic character of the bird names is preserved in Chile until today.

On Chiloe Island, *Williche* populations on the Pacific Coast associate the territorial call of the House Wren, "*yec-yec-yec*," with the sound that is produced when pieces of *rumul kofke* (bread baked in the ashes or "Chilean tortillas") are scratched with the shells of mussels or *quilmahue* are ground. For this reason, on Chiloé the House Wren, the same that the noisy Thorn-Tailed Rayadito, is called "raspa-tortillas" (*Tortilla*-scratcher).

Because the House Wren, or *chedkan*, sings the whole day, the *Lafkenche* call gossipy people *charkanas*, i.e., people who pass their time judging the lives and loves of others more than their own lives. Along the coast of Temuco, *charkan* is also the name of a culinary dish that the *Lafkenche* women prepare with *poñü*, or potato (*Solanum tuberosum*), and *kollof*, or giant kelp (the brown algae, *Macrocystis pyrifera*), because this food has the same color as the noisy *chedkan*.

BIRDS OF THE FOREST MARGINS

Yahgan: **Chílij,** Apuščilix, Hulušénuwa, Apuštöčix, Ūlušönūwa

Mapudungun: **Chedkan**, Chircan, Chedquen, Chedüf, Chüdüf, Chedkeñ, Charkan, Chercan

Spanish: **Chercán,** Ratona común, Raspatortillas

English: House Wren

Scientific: *Troglodytes aedon* (Troglodytidae)

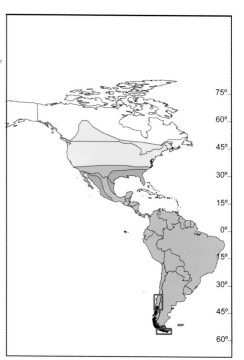

Mapuche Territory ⬜

Yahgan Territory ⬜

Year-round Resident ⚫

Summer Range ⚫

Winter Range ⚫

Resident northern populations; extreme southern populations migrate north during the winter in a variety of forest types
Sighting probability: 4

IDENTIFICATION	4-5" (10-13 cm) Small and faded bird. Brown streaked upperparts, and white underparts with a beige tinge. It has short rounded wings, strong legs, rufous rump, and a cocked rufous barred tail.
HABITAT	It inhabits forests, shrublands, arid and semi-arid environments, parks, gardens and cities. Found from south of Maule to Cape Horn, from the sea level to 3,600 m.
HABITS	Small, active and noisy bird. Alone or in pairs. Territorial and active singer during breeding season. Cavity nester, also using human settlements. It has a fast, varied warble that ends on an accented, often upslurred note. Its raspy territorial song is *"Ratraut, traut, traut, traut, treeeee, ree-traut."*
DIET	It feeds on insects and small terrestrial invertebrates.
CONSERVATION	Thirty known subspecies in the Americas. Until recently all subspecies belonged to the species *Troglodytes aedon*, but today most ornithologists recognize the South American House Wren as a different species: *Troglodytes musculus*. In Chile, the subspecies *T. musculus chilensis* is very common and inhabits from the Atacama to Cape Horn. It is classified as Least Concern by IUCN.

Lantu
Viudita
Patagonian Tyrant

 **CD 1** / Track 33

BIRDS OF THE FOREST MARGINS

The Patagonian Tyrant is a small insectivorous bird, the single representative of the genus *Colorhamphus*, which is endemic to southern South America. During winter, Patagonian Tyrants leave the austral forests, migrating to central Chile or the Neuquen Province in Argentina.

In shrublands,and the edges and interior forests it flies through the lower canopy and perches on the branches of the understory while catching insects. Its small size and its dark grayish-brown color, make this bird difficult to see. Nevertheless, the tyrant broadcasts its presence with its long, sad, descending whistle, *"feeeuuuuu…"*, that peals forth from the foliage like a tragic lament. This whistle reminds one of a widow who walks through the forest grieving for the loss of her husband.

In the Hualakura community, where a stone (*kura*) exists in the form of a *huala* or Great Grebe (*Podiceps major*), there lived a widow who with the same tone as the bird lamented the absence of her dead husband, saying, *"layem fel fotrün, layem fel fotrün"* (Oh, how I desire to have my husband together with me!).

Other communities of southern Chile think that the name of *viudita*, or little widow, derives from its dark grayish color, which resembles the color of the traditional dark, mourning dress worn in Spanish culture by widows. The name of *viudita* has been incorporated into the modern *Mapudungun* name *lantu* (widow), which complements the traditional name given to this bird, *peutren*.

BIRDS OF THE FOREST MARGINS

MAPUCHE STORY

 CD 1/ Track 61

The Mapuche bird name *peutren* means knowing the grandparents and ancestors. Chilean people call this bird "viudita," which means little widow. Perhaps, this name is inspired by the descending pitch of the bird's pitiful call, who sounds like a lady who walks in the forests lamenting her loneliness. In the region of "Nueva Imperial," to the southeast of Temuco, on the shores of the Chol-Chol River, a widow lived who lamented the absence of her dead husband every time she got drunk, and recalled the song of the "viudita" in the forests, when she said, *"layem fel fotrün, layem fel fotrün"* (Oh, how I desire to have my husband together with me, Oh, how I desire to have my husband together with me!).

MAPUDUNGUN VERSION

Tachi peutren (pájaro) kimeluwiki kimtukungeyal pu trem epuñpüle lakuwen ka feyti ka che dungun meu lantu pingey feymay ngüman ülkantu yeniel tachi üñüm, chumngechi ngümayawi kiñe domo rangiñ mawidameu lantungüman. Traytrayko waria püle (Nueva Imperial) lafkenkonün antü püle müley Temüko waria inaleufü choll-choll, mülekerkefuy kiñe lantu domo kiñekemeu pütupütungelu fey may rume duamtukefuy ñi fütrayem laludeuma fey pipingey "layem fel fotrü-layem fel fotrü" ka femngechi may konümpamekekefuy tañi ül tachi lantu rangiñ mawidantu meu.

Yahgan: unknown

Mapudungun: **Lantu**, Peutren

Spanish: **Viudita**, Peutrañ

English: Patagonian Tyrant

Scientific: *Colorhamphus parvirostris*
(Tyrannidae)

Mapuche Territory ☐

Yahgan Territory ☐

Year-round Resident ◯

Summer Range ◯

Winter Range ●

Resident northern populations; southern populations
migrate north during the winter
On a variety of forest types
Sighting probability: 2

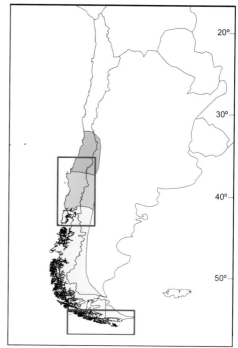

BIRDS OF THE FOREST MARGINS

IDENTIFICATION	4.7-5.5" (12-14 cm) Small bird with gray head, black eyes, and a short and thin, black beak. Upperparts are grayish brown with blackish brown wings with two narrow cinnamon colored stripes. Its belly has light beige color, and its legs are black.
HABITAT	Found in the forest interior, and margins of dense and open forests, as well as shrublands, farmlands, urban parks and gardens. Inhabits from Fray Jorge National Park (30°S), throughout the South American temperate forest biome, down to the Cape Horn Archipelago (56°S); from the coastal to the Andean (2,000m) shrublands and forests.
HABITS	Alone or in pairs. It spends most of its time in the dense canopy of trees and shrubs. Like other tyranids, it likes to perch in branches and catch insects in flight, then return to the same branch. It has a characteristic, long melancholic descending whistle. It is more frequently heard than it is seen. During winter most individuals migrate to the sclerophyllous forests of the central-northern areas in Chile. However, some individuals stay year-round in the sub-Antarctic ecoregion.
DIET	It feeds mainly on insects, but occasionally includes fruits and seeds in its diet.
CONSERVATION	Distribution restricted to the Endemic Bird Areas 060 and 061 (Birdlife International), but is classified as a species of Least Concern (IUCN).

Sámakéar
Pichpich
Cachudito
Tufted Tit-Tyrant

 **CD 1** / Track 34

The Tufted Tit-Tyrant gets its English and its Spanish name (*cachudito*) from its long, black crest feathers that resemble a hairstyle, formerly worn, where the hair would be curled at the nape of the neck (*cachito*).

Together with the Firecrown Hummingbird (*Sephanoides sephaniodes*), the Tufted Tit-Tyrant is the smallest bird in the temperate forests of austral South America. The similarity between the birds did not go unnoticed by the Yahgans, who used the name *sámakéar* for both species: the Tufted Tit-Tyrant and the Firecrown Hummingbird, which are the only austral birds that weigh less than 10 grams.

The Tufted Tit-Tyrant is an anxious bird that moves about constantly between the branches and foliage of trees and bushes, where it feeds on all types of small insects, larvae, and, occasionally, fruits and seeds. From the south of Colombia to Cape Horn, it lives in a wide variety of biomes, utilizing mostly shrubland formations at the edges of forests. Throughout the length of its geographic distribution, it visits human settlements and cities, where it is common in the bushes and trees of gardens and plazas.

The Tufted Tit-Tyrant meticulously builds its nest low in the little branches of the bushes, forming a tiny basket made of grasses, fibers and feathers. From the foliage emerges its trill, which sounds like a smooth chirping in *decrescendo*. It also possesses a territorial song that it usually emits early in the morning that consists of a long sequence of brief, sharp notes.

BIRDS OF THE FOREST MARGINS

Yahgan:	**Sámakéar**
Mapuche:	**Pichpich**, Peshkintun
Spanish:	**Cachudito**, Cachudito pico negro, Torito
English:	Tufted Tit-Tyrant
Scientific:	*Anairetes parulus* (Tyrannidae)

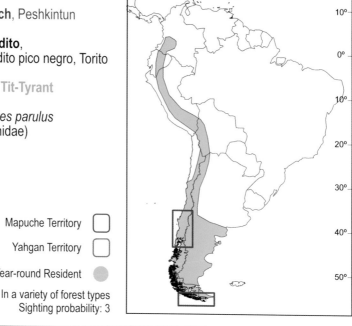

Mapuche Territory ☐

Yahgan Territory ☐

Year-round Resident ⬤

In a variety of forest types
Sighting probability: 3

IDENTIFICATION	4-4.7" (10-12 cm) Very small bird, with a characteristic crest of black feathers that curl forward over the crown. Its head is black with white lines on its forehead, and has an eye-catching white yellowish iris. Its upperparts are dark grayish, while its underparts are yellowish with white streaks on the throat and breast. Its bill and legs are black.
HABITAT	It inhabits a high diversity of habitats, including forest, shrublands, steppes, and farmlands, from the coastal areas to the Andean pre-cordillera (2,000 m). It is also common in urban parks and gardens. The subspecies *Anairetes parulus parulus* inhabits from Antofagasta to Neuquen (Argentina) and Aysen (Chile), and the subspecies *A. parulus lippus* inhabits sub-Antarctic ecoregion, from the Strait of Magellan to Cape Horn.
HABITS	Often solitary or in pairs, but during the non-breeding season can form small groups or mixed flocks with other passerines. Short, smooth and repetitive trill: *phrrrrrrrrr*. Curious, restless but also a shy bird which hides in dense vegetation.
DIET	Feeds on insects, and occasionally includes fruits and seeds in its diet.
CONSERVATION	*Aaniaretes parulus* includes four subspecies, and two of them inhabit the South American temperate forest biome. It is classified as a species of Least Concern (IUCN). Its sister species, *A. masafuera,* is restricted to the Juan Fernandez Archipelago (Endemic Bird Area 059), and is threatened by habitat degradation. **NT**

Chámuj
Chinkol
Chincol
Rufous-Collared Sparrow

 **CD 1** / Track 35

This bird is found from southern Mexico to Cape Horn, where it receives the Yahgan name of *chámuj*. It is characterized by the gray crest it carries over its head. The male's crest is more pronounced and the colors are more marked than the female's. The English name Rufous-Collared Sparrow comes from their chestnut, or rust, colored collar that looks like a scarf. The juveniles lack the tuft and collar.

The *Mapudungun* name *chinkol* means to unite, and denotes the fact that these birds travel in flocks and eat together. This image evokes the collaborative work that expresses the solidarity among *Williche* people and others inhabitants of Chiloe Island, who on occasions like a "minga," or "mingaco,'" carry out a hard work that is accompanied by good food and joyful living together; much like the communal barn raisings that are held traditionally in the United States. Other *Mapudungun* names refer to the characteristic crest that the *chinkol* carries on its head, and that has the form of a hatchet, or *toki*. Likewise, he who carries the axe is the war chief, or *toki*. The Mapuche nation still aspires to unearth the axe of Caupolicán, their greatest leader, an axe that, it is prophesized, could be found by the *chinkol*. Some names that refer to the *toki* of the *chinkol* are: *utruftoki* (throws the axe; *utruf* = throw), *longkotoki* (with the head of an axe; *longko* = head) and *meñkutoki* (has to do with the "hatchet" or tuft over the brown collar that it has on the nape of its neck; *meñku* = carry over the nape or head).

In winter the sparrows form flocks that move, skipping and pecking, across the soil in search of little seeds or insects. As the *Lafkenche* poet, Lorenzo Aillapan says, this bird, or *uñum*, is a "qualified searcher for seeds and crumbs of bread, sometimes with its own kind or in mixed flocks, always in pairs or in big groups" (*kintuyaukey fün fün kofke ka fün ketran, kiñeké meu fentre reyüyaukey kake üñümengün, re mürwen miyaukey newe kiduyaukelay chinkollyaukeyngün*). They are found in almost all environments, although they are more abundant in shrublands and at the edge of forests. In spring, during the reproductive season, they form pairs, and the males frequently emit their melodious and well-known call. In the countryside of central Chile, the *chinkol* is known as "Uncle Agustín," because its call seems to ask, "Have you seen my Uncle Agustín?" On Chiloe Island, when a *chinkol* perches or sings near the entrance to a house, it announces the arrival of a letter or visit of good will.

˙ *"Minga" is a term used on Chiloe Island, which refers to the work done by a group of persons following the request made by a member of the community, who needs, for example, to move or build his/her house, or harvest potato (= poñü).*

159

BIRDS OF THE FOREST MARGINS

Yahgan:	**Chámuj**, Čamux, Kqaiaminix
Mapuche:	**Chinkol**, Chingol, Mencu, Toque, Mencu Lonco, Meñkutoki, Utruftoki, Chincol
Spanish:	**Chincol**, **Chingolo**, Copete
English:	Rufous-Collared Sparrow
Scientific:	*Zonotrichia capensis* (Emberizidae)

Mapuche Territory ☐

Yahgan Territory ☐

Year-round Resident ⬤

Occasional Visitor ⬤

In a variety of forest types
Sighting probability: 4

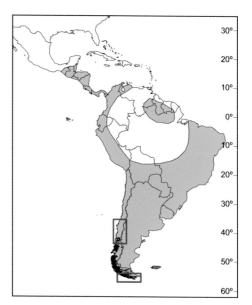

IDENTIFICATION	5-6" (12-16 cm) Males with characteristic erect, gray, crested head. Both males and females have a distinctive broad rufous semi-collar on the nape. Upperparts are grayish brown with blackish streaking; underparts are paler with a light brown breast, and white-grayish belly.Young birds have a duller, indistinct head pattern, with brown stripes and a buff ground color. They lack the rufous collar, and have streaked underparts.
HABITAT	Regarding its habitat use, the Rufous-Collared Sparrow is the most generalist Passerine species of the South American temperate forest biome. It inhabits forests, shrublands, sub-Antarctic tundra, and steppe, as well as open agricultural areas, gardens, parks and side walks in human settlements. Found throughout the South American temperate forest biome, from coastal habitats to the Andean Cordillera (2,500 m). In Chile, the subspecies *Zonotrichia capensis chilensis* inhabits from the Huasco Valley (28°S) to Aysen (45°S), and the subspecies *Z. c. australis* inhabits the sub-Antarctic ecoregion, from Aysen (45°S) to the Cape Horn Archipelago (56°S). In Argentina, *Z. c. australis* in Staaten Island, Tierra del Fuego, and Chubut, and the subspecies *Z. c. choraules* lives in the steppe, shrublands, and forests of Rio Negro, Neuquen, Mendoza, La Pampa, and Cordoba.
HABITS	It perches in tall branches, high rocks or fences, where it is usually seen singing. It is found in pairs or in small flocks and also in mixed flocks. Tame and approachable. The geographical variations of its vocalizations are well documented, and they are known as the "Zonotrichia dialects;" some of them include slurred whistles with a long, final trill, *"tee-teeooo, tre'e'e'e."*
DIET	It feeds on seeds, fallen grains, insects and spiders.
CONSERVATION	Although the Rufous-Collared Sparrow is abundant in most regions, in urban habitats it has been displaced by the House Sparrow (*Passer domesticus*), which was introduced from Europe in 1904. *Zonotrichia capensis* includes 26 subspecies, and three of them are found in the South American temperate forest biome. It is classified as a species of Least Concern (IUCN).

Hashpúl
Püdko
Diucón
Fire-Eyed Diucon

 **CD 1** / Track 36

The call of the *püdko* resembles the falling of water drops or the passing of heavy clouds. The *Lafkenche* communities that inhabit the sector of Lake Budi, associate this call with the coming of a drizzle, because as the poet Lorenzo Aillapan says its name indicates "the *püdko* separates the clouds (*püd*) loaded with water (*ko*)."

The *Williche* communities of Chanquin and Huentemo on Chiloe Island believe that when the *püdko* rises and falls over the same branch, it announces bad weather. These communities call *püdko* by the name *huelko*, who has spiritual powers and is a good shaman that accompanies travelers through the forest, looking out for them from the canopy of the trees and bushes. For other *Williche* communities, *huelko* is a "sent one" that witches dispatch to test the reactions of a person under observation and for this reason, is treated with respect.

The Yahgans also say that one must treat the Fire-Eyed Diucon, or *hashpúl,* well because it is a powerful shaman. It is wise to take care with this "red-eyed one," as it is known on Navarino Island. Grandmother Úrsula Calderón warns that one should never throw stones at the *hashpúl*, because this will bring strong storms with southern winds, or *ilan,* to descend upon them.

YAHGAN STORY

 CD 2/ Track 2

My grandfather said that the *hasphúl* or Fire-Eyed Diucon was a signal of bad fate. If we saw him, we were not to throw any stones after him. If we did so, the *hasphúl* would bring rain, thunder and snow. My brothers did not believe my grandfather, and they threw stones after a *hasphúl* in order to find out whether he was right. So soon afterwards, the bad weather began; it came with rain and thunder. "Wow!" they said. "It is true what the grandfather told us, never again we will do so." Later, my mother confirmed: when the grandfather says something, we have to obey him, because he knows. And surely, it is bad to throw stones after the *hasphúl* bird.

Yahgan:	**Hashpúl**, Héšpul, Ušpul
Mapuche:	**Püdko**, Huelko, Diucon
Spanish:	**Diucón,** Huilco, Hurco, Ojos colorados
English:	Fire-Eyed Diucon
Scientific:	*Xolmis pyrope* (Tyrannidae)

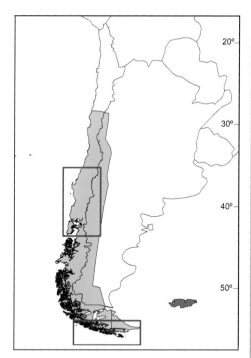

Mapuche Territory

Yahgan Territory

Year-round Resident

Occasional Visitor

In a variety of shrublands, and forest margins
Sighting probability: 3

BIRDS OF THE FOREST MARGINS

IDENTIFICATION	7-10" (19-25 cm) Conspicuous iris of fire-red color in adults, and brown color in immature birds. Its head and upperparts are dark gray, while its throat is white. Its breast and belly are pale gray.
HABITAT	It inhabits forest margins, shrublands, farmlands, and urban parks, from the coastal areas to the Andean pre-cordillera (2,000m), from Neuquen (38°S, Argentina) and Copiapo (27°S, Chile) to Horn Island (56°S).
HABITS	Solitary, in pairs or flocks. Perches on the top of trees, fences or electrical wires, where it emits its characteristic short whistle. Frequently observed feeding on the ground, including in dirt roads feeding on caterpillars and adult insects. It nests on trees or dense bushes, at short distance (1-3 m) above the ground.
DIET	Feeds on adult and larval insects, and occasionally berries and seeds.
CONSERVATION	*Xolmis pyrope* includes two subspecies: the larger and darker one, *X. p. fortis* restricted to Chiloe Island, and the common one, *X. p. pyrope* is found throught the rest of the distribution area. The distribution of the species is restricted to the Endemic Bird Areas 060 and 061 (Birdlife International); yet, it is classified as a species of Least Concern (IUCN).

Tátapuj
Küreu
Tordo
Austral Blackbird

 **CD 1** / Track 37

The Austral Blackbird is a solid black bird that is endemic to Chile and Argentina. In these countries, they say "it's as if the blackbirds were bathed in Chinese ink," for, even the feet and eyes are black. The *Mapudungun* name, *küreu,* could be related to its black (*kurü*) color, as well. Most of the year, blackbirds emit a variety of calls, from strident cries to melodious songs, and form boisterous flocks inside and outside of forests. These vocalizations permit the blackbirds to keep their group together while they feed. The Mapuche name, *küreu,* might also be onomatopoeic with its most common call, "cu-ra-tuuuuu." From the *Mapudungun, küreu,* the Jesuit abbot and naturalist Ignacio Molina derived the scientific name: *Curaeus curaeus.*

The *Williche* people of Chiloe Island attribute distinct meanings to the variable calls of the *küreu.* If they are melodious, or if they cross you in the road and call softly, then your business will go very well, but if its call is harsh your fortune will be adverse. For the *Lafkenche* people, the blackbird possesses an elegant black-blue color, of "very beautiful plumage, dressed in black-blue, shiney, flashy blue of a foreign aspect" (*allangechi tukun kürükallfü pichuntuley, kallpüle tuwün witran rekeley wilüftukun*), that has given him the nickname, "dandy."

Julia González tells that when the Yahgan see the blackbirds, or *tátapuj,* descend to the coasts of Navarino Island and the Archipelago of Cape Horn, it is because snowstorms will come, or because it has already snowed abundantly in the mountains or *tulára.* For the Yahgans and other inhabitants of the extreme southern region, *yakua tátapuj,* or albino blackbirds, who "take the color of the snow" are also well known.

BIRDS OF THE FOREST MARGINS

Yahgan:	**Tátapuj,** Kátapuj, Tetapux, Totupux
Mapuche:	**Küreu,** Cureu, Chihuanco, Quereu, Furare, Kürew
Spanish:	**Tordo**, Tordo patagónico
English:	Austral Blackbird
Scientific:	*Curaeus curaeus* (Icteridae)

Mapuche Territory ☐
Yahgan Territory ☐
Year-round Resident ⬤

In a variety of forest types
Sighting probability: 4

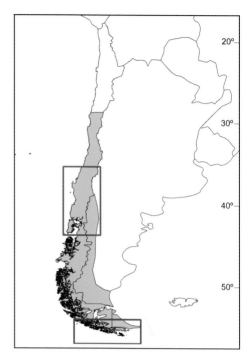

IDENTIFICATION	9.5-11.8" (24-30 cm) Its plumage, bill, and legs are completely black. The female's color is more dull than the male.
HABITAT	It inhabits almost every environment with vegetation, including coastal forests and shrublands, riparian habitats in valleys, farmlands and alpine areas (2,500 m), evergreen and deciduous forests, as well as urban landscapes. Three subspecies are found in central and southern Chile and in south-west Argentina, from coastal habitats to the Andean Cordillera (2,500 m): *Curaeus curaeus curaeus* from Neuquen (36°S, Argentina) and Copiapo (27°S, Chile) to the Strait of Magellan (53°S), *C. c. reynoldsi* from the Strait of Magellan to Cape Horn (56°S), and *C. c. recurvirostris* recorded only on Riesco Island (53°S) in the Magellanic region.
HABITS	Generally in flocks. The members of the flock feed on the ground, or in the middle of tree canopies or shrubs, while one individual watches from a high point. Individuals take turns occupying the role of watch keeper. They are highly vocal with a pleasant song occasionally punctuated by harsher calls.
DIET	Omnivorous bird, feeds on invertebrates, nectar, seeds, and berries. Coastal populations also feed on intertidal marine invertebrates.
CONSERVATION	Its distribution is restricted to the Endemic Bird Areas 060 and 061 (Birdlife International), but is classified as a species of Least Concern (IUCN).

Sámakéär
Pinda
Picaflor chico
Green-Backed Firecrown

 CD 1 / Track 38

Hummingbirds are a bird family found exclusively in the Americas, whose species are concentrated in the tropics. They dart about, and then suspend themselves, seemingly motionless in the air, beating their nearly-invisible wings up to 200 times per second. This extraordinary type of flight is fueled by one of the highest metabolic rates known for any living organism. Because the climate of the tropics is warm, the energy needed to maintain body temperature and fly is less than at cold, high latitudes. For this reason, it is surprising to find a species of hummingbird that reaches the southern extreme of South America in the Chilean Antarctic Province. Moreover, in the austral forests, there are more than ten species of plants, with red, tubular flowers, whose pollination depends on this little bird. As it flies among the flowers, the Green-Backed Firecrown brings an unexpected tropical character to the austral forested landscapes.

The presence of the Green-Backed Firecrown in the austral forests is, however, not only exciting to biologists, but also to poets and lovers. Its Spanish name, *picaflor*, indicates that this little bird, or *pichi uñum*, visits or prick *(picar)* the flowers *(flores)*, and as the *Lafkenche* poet Lorenzo Aillapan expresses:

Its feathers are of very beautiful colors, golden and green and of other colors, and it uses them to fashion gold. At sunset, the fireflies take flight, like lanterns. Their eyes gleam like fire from the nest. By this light the Mapuche women spin, sew, knit and dance, and with one of these nests a girlfriend´s letter can be read. *¡Pin pin pin pin d a a a a pin pin pin pin d a a a a!*	*Pütrümüna azdi ñi pichuñ* *kelü karüg ka karüg ka itrofill kelütuwün ka millatuwe kimfalual* *ella punlu wepümi ñi pelomtuwe kuzdemallo reke* *fey ñi epu kuralnge wilüf wilüfngey relmu reke* *feyti pelomtuwe mew fügwü ka ñizdüfkay kawitral ka püruüyche.* *Feyti pelomtwe mew kiñe üñam ñi wirin chillkatungey.* *¡Pin pin pin pin d a a a a pin pin pin pin d a a a a!*

Its *Mapudungun* names, *pinda, pinguera, pigda* or *piñuda,* are related to the verb *pigudcun* (to rub together), and denotes the sound the wings make during rapid flight. The English name for hummingbirds also alludes to this peculiar humming sound emitted by the wings of the bird. The full English name, Green-Backed Firecrown, literally describes its green back and the males' iridescent, fire-orange crown.

Hummingbirds are admired by many Amerindian cultures, especially the Yahgans. In the austral extreme of the continent, the Green-Backed Firecrown, or *omora,* is an occasional visitor considered by the Yahgan people to be a bird and, at the same time, a small man or spirit who maintains social and ecological order. Today, the figure of *omora* has inspired a conservation initiative that integrates the biological, ecological, anthropological, social and cultural dimensions of life with the goal of promoting the well-being of all species, including humans, in the Cape Horn region. Narratives, such as the following, teach us how *omora* achieves such a life-affirming integration since long ago:

YAHGAN STORY

 CD 2/ Track 1

A long time ago, a great drought occurred in the region of Cape Horn and the people were dying of thirst. Only the wily fox, known as *Cilawáia,* found a lake. He told no one and built, instead, a solid fence around it so that no one could enter. Hidden as such, he drank lots of water, concerned only with himself. However, after some time the other people discovered this lake and went as a group to ask the fox for a little bit of water. *Cilawáia* did not even listen to their supplications and expelled them with rude words.

The condition of the people got increasingly worse and when they were at the point of death, they decided in their desperation to send a message to the hummingbird or little *Omora,* an occasional visitor who had saved them before in times of crisis. *Omora* was always prepared to help others and arrived quickly. Although diminutive, this little tiny bird-man-spirit is braver and bolder than any giant. When he arrived, the downcast people told him about their great sufferings, and, indignant, *Omora* raised himself and rapidly undertook the flight towards where the egotist *Cilawáia* was, and he confronted him directly saying: "Listen! What is it that the people are telling me? Have you had access to a lake full of water, but you don't want to share it with other people who soon will die of thirst!" The fox replied, "Why should I worry myself about the others? This lake has only a little water, and it barely is enough for me and my closest family. I cannot give anything to other people because soon I'll suffer thirst myself."

Upon hearing this, *Omora* became furious and without replying to the fox, returned to the settlement. He reflected briefly and rapidly rose again, took his sling and flew back to the fox. Along the way he collected several sharp stones, and when he was within sight of the fox and sufficiently near, he shouted to him: "Are you going to share your water for now and for always with the other people? Come on, don't be so selfish. If you don't spare water, they surely will die of thirst!"

The fox answered indifferently, "They can all die of thirst. Why should I worry? I can't give water to each one of the people who live here, or else my family and I will also die of thirst." *Omora* was so offended that he could not control himself any longer and furiously fired a sharp stone with his sling, killing the fox with his first throw.

The rest of the community had been watching and became happy, running to the place. Rapidly they broke down the fence, went to the lake and drank, satiating their thirst. They drank so much that the lake became completely empty and a few birds that arrived too late barely found a few drops with which they moistened their throats. Then, the wise owl *Sirra* or *Sita*, the grandmother of *Omora*, said to the birds that arrived late, "collect mud from the bottom of the lake and fly to the peaks of the mountains, and fling the mud over them." The little birds flew, and their balls of mud gave birth to springs that became water courses that spouted from the mountains, forming small streams and large rivers that flowed through the ravines. When all the people saw this, they were extremely happy and drank great quantities of fresh, pure water, that was much better than the water of the lake, and now all find themselves safe again. Today all of those water courses still flow down from the mountains and provide exquisite water. Since that time no one has died of thirst.

Yahgan:	**Sámakéär**, Omora, Samakéa
Mapuche:	**Pinda**, Pinguera, Pinuda, Piñuda, Pigda, Pichi pinda
Spanish:	**Picaflor chico, Colibrí,** Picaflor rubí
English:	Green-Backed Firecrown, Crown Hillstar, Fire Crown
Scientific:	*Sephanoides sephaniodes* (Trochilidae)

Mapuche Territory ☐
Yahgan Territory ☐
Year-round Resident ◉
Occasional Visitor ●
Summer Range ◌
Winter Range ●

Resident northern populations; southern populations migrate north during the winter
In a variety of forest types and shrublands
Sighting probability: 4

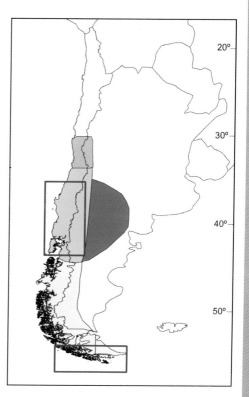

IDENTIFICATION	3.0-4.3" (8-11 cm) A small hummingbird with a large head and short tail. Males have a distinctive metallic, shiny-red forehead and crown. Its upperparts are a shiny metallic-green, with bronze shiny spots. Females are duller, and lack the red crown. Both sexes have white post-ocular spot, and the plumage of their throat, breast, and belly is whitish green, having dark green spots.
HABITAT	It inhabits a variety of forest and shrubland habitats, which include plants with red tubular flowers. It is also found in urban parks and gardens. Found throughout the South American temperate forest biome, from coastal habitats to the Andean Cordillera (2500 m), from Neuquen (36°S) in Argentina and the northern Copiapo Valley (28°S) in Chile to the Cape Horn Archipelago (56°S). It also has populations in the Juan Fernandez Archipelago. It is a common bird in all its distribution range, except south of the Strait of Magellan, where it is an occasional visitor in Tierra del Fuego and Cape Horn. On Navarino Island it has been recorded in summer and early fall, and during the winter southern populations migrate toward the northern areas of Chile and Argentina. The Firecrown has the world's southernmost distribution among all hummingbird species.
HABITS	Alone, in pairs, or small groups. Very territorial, and aggressive with other conspecific individuals. Its flight is characteristically fast and erratic, and while flying emits a characteristic high-pitched, quivering screech. During night it falls in torpor, hanging from twigs with its feet. It builts its nests in high branches or bushes.
DIET	It feeds on the nectar while hovering on the visited flowers, and on flies that it catches during its rapid flights. In the sub-Antarctic ecoregion, especially associated with the red flowers of the Firebush (*Embothrium coccineum*), the Magellanic Fuchsia (*Fuchsia magellanica*), and the Austral Bell flower or coicopihue (*Philesia magallanica*).
CONSERVATION	*Sephanoides* is an endemic genus of south-western South America, which includes two species. *S. sephaniodes* is restricted to the Endemic Bird Areas 059, 060, 061 (Birdlife International), and is classified as a species of Least Concern (IUCN). However, its sister species, *S. fernandensis*, is restricted to Robinson Crusoe Island in the Juan Fernandez Archipelago (Endemic Bird Area 059), and is critically endangered (IUCN). **CR**

Fütra pinda
Picaflor gigante
Giant Hummingbird

 **CD 1** / Track 39

The Giant Hummingbird or *fütra (=big) pinda* is the largest hummingbird of the world. It is surprising to see it beating its wings and emitting its short, sharp whistles in the middle of the forest. Like the Firecrown, it migrates in winter to central Chile. But the migration of *fütra pinda* is much more marked because all individuals leave the austral forests, and some migrate far to the north, since the Giant Hummingbird is distributed through the Andes up to Ecuador. Many rural folks in southern Chile think that during winter, some hummingbirds go into tree cavities and hibernate, to return to the forests with renewed vigor when the flowers bloom in spring. Regarding the hummingbird, the poet Lorenzo Aillapan says:

It nourishes itself with the morning dew, the honey and the liquor of the flowers… Tiny, little bird with so much speed, that while flying, its wings cannot be seen. It sleeps, or nods off, in April, and awakes, or revives, in October, when there are many flowers.	mongekey mülüngmew ka modkoñomew ka llumed pezdkiñmew … kürüf trüri ñi miyawün pichi üñüm üpünüyüm pefalkelay ñi müpü Umagtukey peuwün küyen püle Nepetu mongetukey tromüngen küyem püle

At night, hummingbirds fall into a state of torpor, and during winter, some individuals might seek refuge in holes of trunks and other protected sites for a prolonged torpor period. Among all animal species, these marvelous little birds present the highest metabolic rates, and they still open many new questions to science.

BIRDS OF THE FOREST MARGINS

Hummingbirds not only excite scientists and poets, but also women who want to have children. On Chiloe Island when a woman cannot maintain her pregnancy, it is said that she should trap a hummingbird, caress it in her hands and liberate it immediately. That way, she will receive the powers of fertility that she needs to be a mother.

Several places on Chiloe have *Williche* names that allude to hummingbirds: *Pindapulli*, in the mountains of Pichue, means mountain *(pulli)* of the hummingbirds *(pinda)*; *Pindaco*, water *(co)* of hummingbirds, indicates a leafy place with vines where Chonchi Creek passes; *Pindal*, place of many *pinda* or hummingbirds, lies just to the north of Chonchi. The Giant Hummingbird has left its mark on many places in the Mapuche territory of southern Chile, as well as on aboriginal cultures farther north; it is the same species that inspired the giant hummingbird geoglyph at Nazca, Peru.

Yahgan:	out of bird's range
Mapuche:	**Fütra pinda,** Pinguera, Pinuda, Piñuda, Pinda
Spanish:	**Picaflor gigante**
English:	Giant Hummingbird
Scientific:	*Patagona gigas* (Trochilidae)

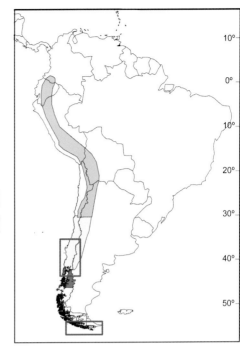

Mapuche Territory

Yahgan Territory

Year-round Resident

Occasional Visitor

Summer Range

Winter visitor to the South American temperate forest biome.
In a variety of forest types and shrublands
Sighting probability: 2

IDENTIFICATION	8-9.5" (20-24 cm) It is the largest member of the hummingbird family. Both sexes have metallic green-gray shiny upperparts with a conspicuous white spot on the rump, and light brown underparts. It has a straight, long, stout bill, and a long, forked tail. The tarsi are feathered to the toes.
HABITAT	*Patagona gigas* includes two subspecies: *P.g. peruviana* that inhabits high Andean puna from northern Chile to southern Colombia, and *P.g. gigas* found from the Huasco Valley (28°S) to Valdivia (40°S), and occasional visitor south of Chiloe, reaching the Guaitecas Archipelago (44°S). In central Chile it inhabits coastal woodlands, xeric open habitats dominated by species of *Puya* (bromeliad plant), and Andean woodlands and shrublands. In southern Chile, it is found in coastal and Andean forests.
HABITS	Alone, in pairs or small groups. Very territorial and aggressive with other bird species. In Central Chile it has a symbiotic relationship with several species of *Puya*, pollinating of its flowers. It almost always chooses to build its nest near the water, especially on top of a horizontal branch of a tree or shrub. The nest is quite small compared to the size of the bird. Its voice is a single loud *"chip"* note. During the austral cold season (March-August), it migrates to northern areas.
DIET	It feeds on the nectar of flowers, and arthropods.
CONSERVATION	*Patagona* is a monotypic genus, and *P. gigas* is classified as a species of Least Concern (IUCN).

BIRDS OF THE FOREST MARGINS

Diwka
Diuca
Common Diuca Finch

(•) **CD 1** / Track 40

"Pewü mew puliwen sollpiwükey diwka": in spring (*pewü*), early morning (*puliwen*), sings (*sollpiwükey*) the Common Diuca Finch (*diwka*). Considered by many birdwatchers to be the loveliest bird-song of Chile, the Diuca Finch beautifies the natural environments and countryside of southern Chile and Argentina with its melodic successions of early-morning whistles: "*diu-diu-diu-diu.*" This song inspires the onomatopoeic *Mapudungun* name, *diwka*, and initiates the morning concert of birds in springtime.

With admirable precision, the song announces the hour of the dawn, which is why when the country people wake up early, they often say they "arise with the first diuca call." It also sings in the harvesting season, when men and women work gathering berries, potato, and more recently, wheat. In the words of the *Lafkenche* poet, Lorenzo Aillapan, the *diwka* "is the bird that only sings in spring and summer seasons, the same as the temporary workers in the countryside, from the end of September to mid March"(*"tachi üñüm re Pewüngen ka Walüngen mütem ülkantukey, feyti pu kona küdeukelu femngechi reke, kon Pewüngen ka epe rangiñ Walüngen"*).

When the *diwka* sings, it is easily visible: usually perching on the canopy of a tree, a protruding branch, a post or a fence wire. It has a characteristic white throat that contrasts with its predominantly gray color. The Common Diuca Finch is distinguished from the Fire-Eyed Diucon (*Xolmis pyrope*) by its black eyes. The adult diucas feed on seeds, and occasionally berries and other fruits from the bushes and forest. The young feed, however, on a variety of fruits, insects and invertebrates. Because it eats mostly seeds as an adult, the Diuca Finch is characteristically called a granivore, yet is really an omnivore.

The Diuca Finch prefers to inhabit prairies, shrublands and other open habitats, but it can also be found at the forest edge, a trait after which the town "Diucalemu" is named. Located in the region of the Bio-Bio River, the Spanish name of this town derives from the previous *Mapudungun* name: *Diwkalemu,* which means forest (*lemu*) of the *diwka*.

Yahgan: out of bird's range

Mapuche: **Diwka,** Shiwka, Fiuca, Diwka, Viuca, Diuca

Spanish: **Diuca**

English: Common Diuca Finch

Scientific: *Diuca diuca* (Emberizidae)

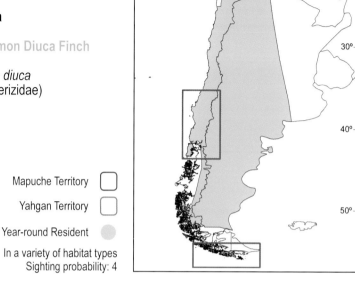

Mapuche Territory ☐

Yahgan Territory ☐

Year-round Resident ●

In a variety of habitat types
Sighting probability: 4

BIRDS OF THE FOREST MARGINS

IDENTIFICATION	5.5-7" (14-18 cm) Gray head, neck, upperparts and chest. The throat and belly are white with a characteristic inverted V-shape entering the lower breast. Dark, short and curved bill. The female is paler.
HABITAT	It lives in a variety of habitats such as open forests, sclerophyllous forests, shrublands, farmlands, as well as rural towns, parks, and urban gardens and parks. From coastal areas to the Andean pre-cordillera up to 2,000 m. *Diuca diuca* has four subspecies, of which two are found the region of the temperate forest biome: *D.d. diuca* inhabits from Neuquen (36°S) to Santa Cruz (53°S) in Argentina, and from Coquimbo (30°S) to the area of Puerto Natales (52°S) in Chile, and occasionally in Tierra del Fuego; *D.d. chiloensis* restricted to Chiloe Island.
HABITS	Alone or in pairs, however it lives in mixed flocks in winter. The male early morning song is a soft and musical *"tiiup tweep chwuup TWEEIP."* It feeds on the ground, but is often seen perching on shrubs, trees or fences. Its nest is made of grasses and fibers, and it is usually placed on a shrub or small tree.
DIET	Feeds mainly on seeds, but its diet also includes fruits, insects and other invertebrates.
CONSERVATION	The distribution of *Diuca diuca* is restricted to southern Argentina and Chile, but is classified as a species of Least Concern (IUCN).

Wichóa
Lloyka
Loica
Long-Tailed Meadowlark

 **CD 1** / Track 41

The poet Lorenzo Aillapan says that the Mapudungun name, *lloyka,* comes from the terms *llako* (to heal) and *lawen* (remedy). According to this Mapuche Bird-Man, or *Uñumche*, the *lloyka* is a healer-bird that, like the *machi* (Mapuche shaman), gets its medicines from the plants of the forests:

Have no fear Great Father and Great Mother, I will have the remedy in time, abundant and good. For this, I call myself another red-breasted being. I have always healed my people with pure herbs. For this, my name is *full-time healer.*	*Llükalayaymi turpu fütra chaw ka eymi kude ñuke Inche ta fey lawentuayu feula fachante frentren ka tutelu fey mew ta llapüm chefe kelü pütra pingeken Ka fey rumel llakotukey pu che re mapu lawen mütem Inche ta fey mew mongelchefe rumel lloyka pingen**

In central Chile, it is said that the *lloyka* call expresses, "with a knife it was, with a knife it was," alluding to the moment when the Long-Tailed Meadowlark thrust a knife in its chest, which shines today, as if it was still bleeding, covered in red. Some Mapuche say that perhaps this *uñüm* (bird) still desires to heal its chest, which carries the blood of the Mapuche nation and the flower of the *kolkopiw* (*Lapageria rosea*).

In the Yahgan territory at the austral tip of the continent, Long-Tailed Meadowlarks are abundant only along the northern coastline of the Beagle Channel. South of that, they are very scarce. The grandmothers Úrsula and Cristina Calderón remembered the name Long-Tailed Meadowlarks or Red-Breasted Starlings from their young years when they lived on the lands of the Anglican missionaries at Harberton Ranch in southern Tierra del Fuego, Argentina. However, after so much time and being infused with other notions of the bird, it took more than one year, working in the field together, before

Grandmother Úrsula could remember the Yahgan name, *wichoa*. Maybe this healer bird, *wichoa* or *lloyka*, can heal our memory at the austral tip of South America, and communicates to us the importance of knowing, watching over and valuing the diverse bird species, and appreciating the diverse cultures that inhabit the austral forests.

MAPUCHE STORY

 CD 1/ Track 6

The *lloyka* or Long-Tailed Meadowlark is a healer bird who transports medicines from one place to another. Among the medicinal plants the *lloyka* transports are pieces of a sacred tree named *foye* (*Drimys winteri*). These pieces are applied to the *rehue*, carved tree trunk posted outside the house of the shaman, who is known as the *machi*. The *rehue* symbolizes the powers of the shaman, and like her house or *ruka*, faces the northeast where the sun or *antü* rises in the morning.

MAPUDUNGUN VERSION

*"Wichin wichin wichin wichin wichin küpalu piam pu lonko Trapuuu
wichin wichin küpalu piam lantuuu kangey trekan lantuuu!"*
Tachi lloyka llakonchefe üñüm yekey fill lawen fill püle. Tachi lawen foye pingelu lloyka ta yey muñküpüle fey wefi foye (*Drimys winteri*). Allangechi lawen aliwen fey kuyfimel rewengekey, kiñe mamüll foye feyti papay machi kintuy tukuy tañi rewe wülngiñ ruka püle, cheutañi tripan antü. machi, dulliñ, lawentufe domo tuwükülelu wüne newen püllümeu llapümkelu kutran kafeyti fill weda neyen kutran).

Yahgan: **Wichóa**, Wičõa

Mapudungun: **Lloyka**, Loica, Lloica, Loyka

Spanish: **Loica**, Pecho colorado, Loica común

English: Long-Tailed Meadowlark, Red-Breasted Starling

Scientific: *Sturnella loyca* (Icteridae)

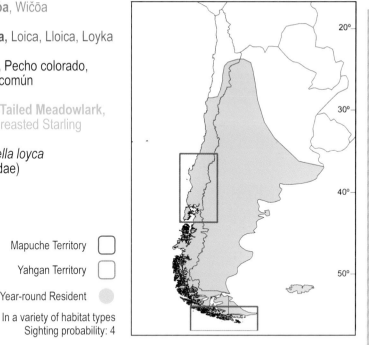

Mapuche Territory ☐

Yahgan Territory ☐

Year-round Resident ●

In a variety of habitat types
Sighting probability: 4

IDENTIFICATION	8.7-11" (22-28 cm) Bird characterized by its bright red breast, throat and upper belly. Its lower belly is dark gray, and its upperparts are black brownish. The head is black with a white superciliary line, and red lore. The female is paler. Black, long and pointed bill with bluish jaw.
HABITAT	It prefers open habitats, such as savannah, open forests, shrublands, grasslands, ravines with dense bush cover, Patagonian steppe, wetlands, farmlands, and also urban parks. From coastal areas to the Andean cordillera (2,500 m). It has four subspecies inhabiting Chile and Argentina, of which one is found in the region of the temperate forest biome: *Sturnella loyca loyca*, which inhabits from Neuquen (36°S) to Tierra del Fuego (55°S) in Argentina, and from Copiapo (28°S) to Navarino Island (55°S) in Chile.
HABITS	In pairs, but forms flocks during winter. It is usually seen perching on the top of bushes or fences, making it possible to distinguish the bright red chest of the Meadowlark from a long distance. Its song is a sharp series of notes followed by a nasal whine: "*chip-kik-cut-cheo ZZCHWEEEEEOOO.*" The nest is made of dry grasses, placed on or near the ground, where it also feeds.
DIET	It has an omnivorous diet, which includes seeds, fruits, insects and other invertebrates, such as worms, and snails.
CONSERVATION	The distribution of *Sturnella loyca* is restricted to Argentina and Chile, but is classified as a species of Least Concern (IUCN).

Kamtrü

Rara

Rufous-Tailed Plantcutter

(•) **CD 1** / Track 42

Among the birds inhabiting the temperate forests of South America, the Rufous-Tailed Plantcutter is the only species that is primarily herbivorous. There are very few, small herbivorous bird species in the world. Most herbivorous birds weigh more than 1 kg and browse on grasslands, such as ostriches and geese. The Rufous-Tailed Plantcutter, in contrast, flies among trees and shrubs eating the flowers, buds, and the tender leaves of several plant species, with its finely serrated bill.

The male has distinctive copper colored undertail coverts, and on the belly, throat, and crown. Females are duller, but both sexes have bright red eyes. Finally, the most peculiar characteristic of this bird is the unmistakable territorial call that the male utters while perched on a high tree branch. This call consists of a sequence of rasping sounds that Chilean ornithologist Guillermo Egli compares with the sound of a matraca-rattle.

It is precisely from this strange, percussive call of the Rufous-Tailed Plantcutter (*ke-ke-ke-ke-kreeé* o *rara-rara-rara*) that the onomatopoeic *Mapudungun* names *kamtrü* and *rara* derive. This call is heard as much in the forests of the mountains of the *Pewenche* territory, as it is on the coastal habitats of the *Lafkenche*. *Kamtrü* lives not only in the forests or *mawida*, but also in the agricultural lands, or *ngan*. Lorenzo Aillapan explains that, on the coast of Temuco the Rufous-Tailed Plantcutter has a special preference for broad bean plantations. For this reason, this bird is known locally as *cortaplantas* (plantcutter). However *kamtrü* are not harmful for crops, and due to its low numbers, today, its hunting and capture are prohibited.

Yahgan: out of bird's range

Mapudungun: **Kamtrü,** Rara, Cadi

Spanish: **Rara**

English: Rufous-Tailed Plantcutter

Scientific: *Phytotoma rara*
(Cotingidae)

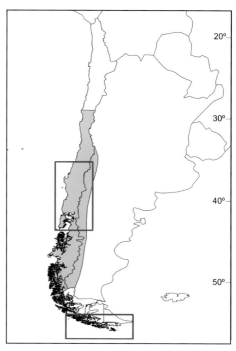

Mapuche Territory ▢

Yahgan Territory ▢

Year-round Resident ⬤

Southern and high-altitude populations migrate
northwards or towards the lowlands in winter
In a variety of forest types and shrublands
Sighting probability: 2

BIRDS OF THE FOREST MARGINS

IDENTIFICATION	7-8" (18-20 cm) Forehead, crown, throat, breast, and belly with a distinctive rufous copper color. Upperparts are blackish brown, with a conspicuous white line on its blackish wings. Females are more dull. Both sexes have a salient, red iris. Its strong, short, black beak is also peculiar: it has serrated edges specialized for eating plants.
HABITAT	Inhabits sclerophyllous forests, savannah, farmlands, orchards, as well as the margins of evergreen and deciduous rainforests, from coastal to Andean habitats (2,500m) in southern Chile and Argentina. Occasional visitor in urban gardens and parks. It is found in western Argentina from Mendoza (33°S) to the area of Perito Moreno Calafate (50°S), and throughout Chile from Copiapo (28°S) to the area of Puerto Natales (52°S); occasional vistor to the forests of Tierra del Fuego (54°S).
HABITS	In pairs most of the year, during fall and winter forms small flocks. Its serrated bill is used for stripping off buds, leaves and fruits. Its characteristic vocalization consists of a series of stuttering notes followed by a rasping trill similar to a fishing reel (*ke-ke-ke-ke-kreee*).
DIET	The only herbivorous (folivorous or leaf-eater) bird species in the South American temperate forest biome. It also feeds on buds, and fruits, and occasionally on insects.
CONSERVATION	Distribution restricted to the Endemic Bird Areas 060 and 061 (Birdlife International), but is classified as a species of Least Concern (IUCN). In 1902 the book "La Caza en el Pais" (Hunting Regulations in Chile by Federico Alberts) included *Phytotoma rara* among the birds species that were considered harmful for agriculture. Today, the Chilean Hunting Laws and Regulations considers this species to be "beneficial for the equilibrium of ecosystems," and as having low population densities. Therefore, its hunting is forbidden. To conserve this species in the long-term, it is necessary to conduct programs communicating the new understanding of its ecological value, and the new hunting regulations.

RAPTORS
OF THE FORESTS AND ADJACENT HABITATS

RAPTORS

Yoskalía
Triwkü
Tiuque
Chimango Caracara

 **CD 1** / Track 43

The loud descending cries of the Chimango Caracara, "*triiiiuuuu, triu, triu, triu, triu ....*", characterize the forests and other environments of southern Chile, and are the source of its onomatopoeic *Mapudungun* name: *triuki.*

This is the most common raptor in and around the austral forests. The Chimango Caracara uses trees to sleep and nest, and builds its large nests with twigs and branches. Although it is mainly a scavenger, it is omnivorous, as well, hunting frogs, lizards, mice, small fish, insects, earthworms, larvae, caterpillars and even slugs. Therefore, it is a very beneficial bird for agriculture. When cows are browsing in the prairies or people are hoeing the land, groups of up to one hundred Chimango Caracara are seen eating insects on the ground.

The *Williche* on Chiloé Island consider the Chimango Caracara or *triuki* a "suspicious bird" because witches use them and even transform themselves into them. When the Chimango Caracara lands on the roof of a house, is said that it may be a witch that listens to the conversations of the people inside. For the *Lafkenche*, in contrast, the *triuki* brings happiness by accompanying them during agricultural labors and by removing the lice from the animals. In addition, they help the farmers because their cries call rain to soak the land, and they eat grubs and other crop pests.

MAPUCHE STORY

CD 1/ Track 64

The cry of the *triukü, triiiiuuuu, triu, triu, triu, triu...* invokes water, humidity and rain, because they feed on worms and grubs in the moist earth. Consequently, it is a beneficial bird for farmers.

MAPUDUNGUN VERSION

Feytachi pin dungun triuki trüüuuuu, triu, triu, triu triu, mütrümi komawün, ka fochon ka feyti mawün, feykam ikelu düllwi, katachiwera piru fochon tuwe meu. Feymeu tati, küme üñüm kellukefi pu küdaufe fünketrankelu.

RAPTORS

Yahgan:	**Yoskalía,** Yookalía, Yōakölia
Mapudungun:	**Triwkü,** Triuki, Tiuque, Chiuque, Triuquem, Chiwkü
Spanish:	**Tiuque,** Chimango
English:	Chimango Caracara, Tiuque
Scientific:	*Milvago chimango* (Falconidae)

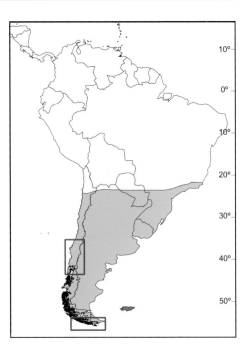

Mapuche Territory ☐

Yahgan Territory ☐

Year-round Resident ●

Occasional Visitor ●

In a variety of forest types
Sighting probability: 4

IDENTIFICATION	14.6-15.8" (37-40 cm) Medium sized and slim bird. The Caracara has a distinguishing brown color. Upperparts are dark brown, while lower parts are pale, cinnamon brown, and the throat is even paler. During flight, its long wings exhibit a conspicuous white patch on the base of the primary feathers, and its long, pale tail has a dark brown subterminal band, and a characteristic white, terminal band. Head and crown dark brown, ivory colored beak with bluish base, and yellow cere*.
HABITAT	It inhabits a wide range of habitats, including temperate and sub-Antarctic forests, coastal habitats, shrublands, grasslands, riparian meadows, and wetlands, from the intertidal to the Andean pre-cordillera (2,500m). It also frequents human settlements, from towns, and farmlands, to large cities, and dump sites. *Milvago chimango* includes three subspecies, all of which are found in the South American temperate forest biome: *M. c. chimango* inhabits Chile from Chañaral (27°S) to Concepcion (37°S), central and northern Argentina, Uruguay, Paraguay, southern Brazil and Bolivia; *M. c. temucoenis* lives from Concepcion (37°S) to the Strait of Magellan (53°S) in Chile, and from western Neuquen to Santa Cruz in Argentina; *M.c. fuegensis* is restricted to Tierra del Fuego, and the area of the Cape Horn Biosphere Reserve (56°S).
HABITS	Alone, in pairs, or in small or large noisy flocks. They perch in large trees, especially at dusk, when they return to sleep. During the day they fly around their territories feeding mostly on the ground, including places near human settlements, such as fisheries, farmlands, and dumps.
DIET	Scavenger, but it also hunts small vertebrates, scratches the ground looking for larvae, and all kinds of terrestrial, freshwater, and marine invertebrates, and eats eggs, fruits and seeds.
CONSERVATION	*Milvago chimango* lives only in the southern cone of South America and includes three subspecies. It is classified as a species of Least Concern (IUCN).

*The cere of raptors is the skin surrounding the nostrils and the base of the upper jaw of the beak.

RAPTORS

Ketéla
Traru
Traro
Southern-Crested Caracara

CD 1 / Track 44

The presence of the Southern-Crested Caracara is often detected by its loud squawk, "*K-a-r-a-k, K-a-r-a-k*, or *traru-traru-traru...,*" which is heard from long distances away. These calls are echoed in the onomatopeic *Mapudungun* name, *traru*, as well as in its Spanish translation, *traro*. It is noteworthy that its English name, Caracara, also imitates this bird's vocalization, which was first translated into a name by the Tupi indigenous people in southern Brazil. In turn, in 1826 the German ornithologist Blasius Merrem used this bird's Tupi indigenous name, to create the genus "Caracara" within the Falconidae family. Today, the common name caracara is widely used not only by English speakers, but also by Portuguese and Spanish speakers, mainly in southern Brazil and northern Argentina.

The onomatopeic *Mapudungun* name, *traru*, is still widely used in southern Chile and Argentina. When it calls, the caracara characteristically throws its head backward, displaying its breast, neck, and crest. Due to the prominent crest that it carries over its head, the *Lafkenche* people compare the *traru* with the "carabineros" or Chilean national policemen, who are figuratively called *longko-traru*, because, like the *traru*, the "carabineros" wear an elegant police cap on their heads, or *longko*.

The Southern-Crested Caracara is a typical bird of prey and scavenger in the austral forests and other environments of South America. In 1999, scientists recognized it as a distinct species from its sister species, the Caracara of Central and North America *(Caracara cheriway)*. However, their distinctive vocalizations and the black crests over their heads are characteristic of both species. The Southern-Crested Caracara builds extraordinarily large platform nests, made of twigs and branches covered with soft material, in the canopies of tall, old trees. For this reason, conservation of the forests is very important for the survival of this species. Southern-Crested Caracara is a year-round resident, and pairs will stay in the same territory for many years. Young birds must travel many kilometers to establish their new territories. Omnivores, they eat small mammals, lizards, frogs, insects, crabs, worms and carrion. Occasionally, they attack lambs. The latter image should not diminish the positive contributions that Southern-Crested Caracaras make by "cleaning" the countryside of pests and controlling plagues.

RAPTORS

The Mapuche people have great admiration for raptors and the *traru* has a special historical significance to their culture. The lineage of eagles and Caracaras gave rise to the greatest of all Mapuche warriors: *Leftraru*, whose name means quick (*lef*) *traru*. Known by the Spaniards as Lautaro (1535-1557), *Leftraru* was a superb war strategist, organizing his people into distinct flanks of attack in a way that shocked and startled the Spaniards. *Leftraru's* martial skills were complemented by his profound knowledge of the topography of his land, and by his ability to predict climatic phenomena. His deep understanding of the Mapuche territory allowed him to defeat the Spanish conquistador Pedro de Valdivia. The war strategies of this valiant and intelligent warrior have not only inspired generations of Mapuche people, but also, as poet Lorenzo Aillapan points out, Napoleon was inspired by the tactics of *Leftraru,* which helped him to confuse the enemy and attack it simultaneously from different fronts.

The name of *Leftraru's* father was *Kürüñamku*, which means black (*kürü*) eagle (*ñamku*). As is the case of many Mapuche families' ancestry, the lineage of this hero's family is intimately linked to the swift raptor birds, or *norche*: people (*che*) who love justice (*nor* = rights) and liberty (*naytun* = liberty). Unfortunately, the clear-cutting of native trees in the *Lafkenche* territory has eliminated the places where this noble bird usually nests, and its populations are diminishing drastically. Our goal is to contribute to the protection of these coastal, native forests and to preserve the Crested Caracara and the values inherited from *Leftraru.*

MAPUCHE STORY

 CD 1/ Track 65

Leftraru means quick *traru* or Caracara. This was the *Mapudungun* name of the great Mapuche warrior-hero, known as Lautaro, whose father was called *Kürüñamku*, which means black eagle. So, the family lineage of this hero, like that of many Mapuche families, is intimately tied to birds.

MAPUDUNGUN VERSION

Tachi Leftraru feyta wüne fütra aukafe mapuche ngey, kimngetuy Lautaro üyngen, ka feytañi chau Kürüñamku pingefuy (águila negra). Femngechi, kom küpal lakuy wüne ñidolaukafe, kafey inapalu pu mapuche, üñüm meu tuwükuley laku küpal.

RAPTORS

Yahgan:	Ketéla
Mapudungun:	**Traru**, Traro, Taru
Spanish:	**Traro**, **Carancho**, Huarro, Caracara
English:	Southern-Crested Caracara
Scientific:	*Caracara plancus* (Falconidae)

Mapuche Territory ☐

Yahgan Territory ☐

Year-round Resident ⬤

In a variety of forest types and open habitats
Sighting probability: 3

IDENTIFICATION	19-25.6" (49-65 cm) It is the largest caracara in South American temperate forests, and has a distinctive black-brown crown with a small erect crest. Strong, yellow and/or bluish beak with a reddish orange cere˙. Its neck sides and throat are beige, its breast and upperparts are light brown with dark brown bands. Its wings are dark brown, and legs yellow. In flight, large conspicuous whitish buff patches are visable on the outer primary feathers.
HABITAT	It inhabits from the coast to the Andean pre-cordillera (2,000 m), in a large diversity of habitats, including forest borders, shrublands and grasslands. It is a typical bird in the sub-Antarctic forests, Patagonian steppes associated with sheep ranches, and dump sites. *Caracara plancus* includes four subspecies, one of them, *C. p. plancus*, inhabits Chile from Arica to Diego Ramirez archipelago, and Argentina from Buenos Aires to Staten Island and the Malvinas Islands.
HABITS	Alone, pairs, or in groups, especially during winter, and when feeding. Often seen walking around on the ground looking for food, but it perches and nests in the canopy of trees. Its nest is a large open structure made of twigs, grass, and feathers, placed on the top of a tall tree, but occasionally atop shrubs, cliffs, or rocks.
DIET	Scavenger, but also hunts small vertebrates, insects, and other invertebrates, including marine invertebrates on accumulations of algae beached on the coast. Its hunting behavior is different from other raptors; instead of throwing itself on the prey, it lands at a certain distance and then walks toward it and chases the prey until trapping it. Farmers consider this bird damaging for cattle and domestic bird breeders.
CONSERVATION	Very rare in its northern distribution in Chile, but very common in the south. It is classified as a species of Least Concern (IUCN).

˙The cere of raptors is the skin surrounding the nostrils and the base of the upper jaw of the beak.

RAPTORS

Kīkinteka

Okori

Cernícalo

American Kestrel

(·) **CD 1** / Track 45

The American Kestrel is a species of small falcon that is abundant throughout the Americas with the exception of the Amazon rainforests, the Arctic tundra and boreal regions. In the forests of southern South America, the *Mapudungun* names given to this little falcon suggest the three cognitive dimensions from which the Mapuche culture names the birds: 1) onomatopoeia, 2) cultural-spiritual, and 3) natural history:

1) The name, *okori,* describes the sound of the whistle the kestrel makes when it imitates the hiss of its snake or reptile prey. Among the Mapuche, the word *mutrir* indicates the most intense hiss of a snake, or *filu,* which is also the warning cry that one calls to another. This high-pitched hiss, emitted by the *filu* or by *okori,* can harm the ears of humans.

2) The name *lilpillañ* expresses the cosmology in which the kestrel and other raptors acquire the character of guardian (*pillañ*) of the hills and volcanoes with their cliffs and rocks (*lil*). The American Kestrel or *lilpillañ* inhabits the mountain slopes, occupying a variety of environments. *Lilpillañ* is also a skillful flyer, and during the flights makes a series of strong, short cries, which give it the character of a protector bird. In the mountains, it nests in all types of natural cavities of rocks, cliffs and in the austral forests it takes advantage of the hollows excavated by woodpeckers in large trees.

3) The name *Lleylleykeñ* comes from a natural history observation and refers to the kestrel's habit of suspending itself in the air to observe its prey. The stationary-soaring kestrel concentrates its senses, listening to sounds and using its intensely perceptive eyes to spot the slightest movement of prey below. This disposition is similar to a state of being quiet and concentrated with alert senses, which the Mapuche call *lleylle* state. For this reason, it is called *lleylle;* the suffix *keñ* is applied to the name, a noun, which describes how the bird behaves. *Lleylleykeñ* performs its stationary flight mainly on open lands where it hunts rodents, little birds, reptiles, and a large variety of insects.

191

RAPTORS

Yahgan: **Kĩkinteka**, Akĩmakaia

Mapudungun: **Okori,** Ocori, Lleullequen, Lilpillañ, Lleylleykeñ

Spanish: **Cernícalo,** Halconcito común, Halconcito colorado, Quechi-quechi

English: American Kestrel, Sparrow Hawk

Scientific: *Falco sparverius* (Falconidae)

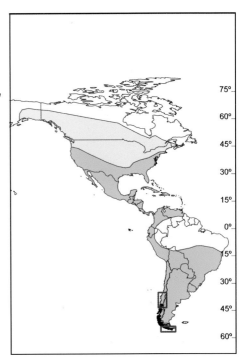

Mapuche Territory ☐

Yahgan Territory ☐

Year-round Resident ●

Summer Range ●

Common in diverse forest types, and shrublands
Sighting probability: 4

IDENTIFICATION	9.8–11.8" (25-30 cm) Smallest falcon of the Americas, with a distinctive gray bluish crown, with a brown reddish patch on the central back part. Sides of the head are white, with two narrow, vertical black facial markings descending before and behind the eye. Short, bluish beak with a yellow-orange cere*. Underparts are whitish with black spots, and upperparts are brown reddish with black barring. Yellow legs. Wings gray bluish with black spots, and while perched, the wingtips are noticeably shorter than the tail tip. The male's tails have one black subterminal band, while female's tails have several black bands.
HABITAT	It inhabits from the coast to the high Andean cordillera (3,800 m), on a variety of humid and xeric habitats, including wetlands, prairies, shrublands, the margin of dense forests, alpine zones, farmlands, and cities. *Falco sparverius* inhabits all the Americas, from western Alaska to Diego Ramirez Islands (56°30'S).One subspecies, *F. s. cinnamominus,* is the only one present in the Chilean and Argentinean temperate forest biome.
HABITS	Solitary or in pairs. This kestrel is a skillful and silent hunter. From elevated perch sites, it waits for prey to move on the ground, or during flight, it can also remain static in the air, flapping its wings quickly when a prey is located. It nests in holes in trees, cliffs, or buildings.
DIET	It feeds on birds, and other small vertebrates, as well as on a wide array of invertebrates, including grasshoppers and beetles.
CONSERVATION	*Falco sparverius* has an extremely large range, and it is classified as a species of Least Concern (IUCN).

RAPTORS

Kíkintéka
Kalkiñ
Águila
Black-Chested Buzzard-Eagle

CD 1 / Track 46

The prefix of the Mapudungun name for the Black-Chested Buzzard-Eagle (*kalkiñ*) means "comes from another place," and alludes to the fact that this bird comes from remote lands. It is a notable visitor for whom the Mapuche have great admiration. The Black-Chested Buzzard-Eagle is the largest raptor species in the temperate forests of South America. Its range extends throughout the Andes: from north to south—Mérida, Venezuela to Cape Horn—and from east to west—the forest south of the Chaco in Paraguay and the xeric shrublands of Córdoba, Argentina to the Pacific Ocean.

The Black-Chested Buzzard-Eagle is seen from far away flying in circles at high altitude and, from close at hand, perching over tall trees at the edge of the forest. It builds its nest of sticks over the canopy of tall trees or on rocky cliffs, and it occupies this nest for several years. It feeds on a variety of medium-sized prey, including rabbits, hares, rats and also, in the extreme south, the exotic, invasive beavers and muskrats. Therefore, it is a beneficial species for controlling the numbers of rodent and small mammal populations.

In flight, the Black-Chested Buzzard-Eagle presents an unmistakable triangular shaped silhouette; the base of its wings is very wide and the tail has a wedge shape. Its upper parts are gray and black, in contrast to the underside, which is white, except for the characteristic "black shield" on its chest that provides the source of its English names, Black-Chested Buzzard-Eagle or Shield Chested Eagle.

In the ancient stories recorded in the territory of the *Pikunche*, the white abdominal feathers of the eagle are of central importance. In the area of Vilches, towards the mountains near the city of Talca, it is believed that, in ancestral times, the eagle saved humanity. When there was a great downpour

RAPTORS

caused by *Kai-kai vilu**, the people went up the mountain where *ten-ten vilu** lived. As the floodwaters rose, the people climbed higher and higher up the mountain. When they climbed as high as they could go and were still unable to escape the deluge, the eagle appeared from the heights of the Andes and with one flick of its beak opened a crater in the volcano. There, the ancestors of Mapuche took refuge. Until this very day, the volcano remains covered with the snows, which are the white feathers of this eagle who saved the Mapuche people.

The story of kai-kai and ten-ten presents many variations in different regions of the Mapuche territory, and at different historical times. The version presented here was recorded in the area of Altos de Vilches (VII Region, Chile) in 1994.

Yahgan:	**Kíkintéka**, Kíkentéka
Mapudungun:	**Kalkiñ**, Caltchi, Calquin
Spanish:	**Águila**, Aguila mora, Aguilucho grande
English:	Black-Chested Buzzard-Eagle, Shield Chested Eagle
Scientific:	*Geranoaetus melanoleucus* (Accipitridae)

Mapuche Territory ☐

Yahgan Territory ☐

Year-round Resident ⬤

In a variety of forest types
Sighting probability: 2

IDENTIFICATION	23.6-30" (60-76 cm) The largest Accipitridae species in Chile, with characteristic soaring flights that expose its unmistakable silhoutte of broad wings and short tail. Head and upperparts are dark gray, cheeks and throat are paler, blackish breast, whitish belly. Females are larger than males, and juveniles have dark brown plumage for 3 to 4 years.
HABITAT	From coastal to mountainous areas (4,000 m) with sparse vegetation, shrublands or temperate forests in its southernmost distribution. *Geranoaetus melanoleucus* includes two subspecies. Only *G. m. australis* inhabits the temperate forest biome; it is found in Chile from Arica to Cape Horn (56°S), and from north western Argentina to Tierra del Fuego.
HABITS	Solitary or in pairs, but in groups during breeding season. Frequently seen soaring. Generally silent, except when it feels its nest is threathened, or when flying, it occasionally emits a high pitched vocalization: *"ku-ku-ku-ku-ku."* It nests in high trees or on rocky cliffs, but also in bush or even on the ground.
DIET	Hunts medium sized mammals, including introduced European rabbits and hares, as well as native mammals and birds, such as owls, geese and tinamous, invertebrates and carrion.
CONSERVATION	It has a wide Andean and South American distribution, and is classified as a species of Least Concern (IUCN).

RAPTORS

Ñamku
Aguilucho
Red-Backed Hawk

(•) **CD 1** / Track 47

The *Mapudungun* name of the Red-Backed Hawk means "departed spirit" (*ñam*) which eventually returns to loved ones, and is of good omen. The suffix *ku* makes the name a noun. The Red-Backed Hawk flies high above the forests of South America, detecting its prey with sharp eyesight and launches itself to capture them. Its plumage is found in many varieties, but the characteristic plumage of the adult can be recognized even when it flies high overhead; the white colors on the front and the black edges of the wings and tail always stand out. It usually perches on rocks in the mountains and exterior branches of large trees in the austral forests. In the south, it constructs nests from branches on tall trees, from where its loud, piercing cry can be heard. It feeds on rodents and birds, but also consumes reptiles like snakes and lizards, and occasionally, even insects such as beetles. Inhabiting all types of environments, including arid zones and rainforests and from sea level to 4,500 m, it is found throughout the Andes from the Sierra de Santa Marta in Colombia to Cape Horn, Chile.

Within the Mapuche territory many people liked the name of this bird, and called themselves *Ñamkupil*, which means "in the spirit (*pil* or *püllü*) of the Red-Backed Hawk (*ñamku*)"; or, *Ñamkulef* which means "swift (*lef*) *ñamku*"; or *Ñamkuan* which means "*ñamku* of the sun (*an* or *antü*)"; and, *Ñamkucheo* which indicates that the person has a great likeness (*cheo*) to the *ñamku* or Red-Backed Hawk. In addition, for the Mapuche people, *Ñamku* is the guardian of the herds, and is respected and always welcomed. The deep empathy between the people and *ñamku* is expressed not only in names and habits, but also in ancestral stories, such as the following:

MAPUCHE STORY

 CD 1/ Track 66

Many people carry a feather of the *ñamku* as a keepsake so that things always go well. Sometimes, the *ñamku* removes a feather and allows it to fall. This happened to the young man, *Mankian*, who, after collecting the feather that the *ñamku* had dropped, became the Sea Guardian, called *Dumpall* or Guardian of the West.

Dumpall accompanies the Volcano Guardian of the Southeast named *Pillan*, the Princess of the Sun from the Northeast named *Anchümallen*, and the Great Visitor who has the treasures from the North, *Witranalwe*.

MAPUDUNGUN VERSION

Fentren che yeniey *ñamku pichuñ* rumel konumpa nieyal fey ñi kümelkayawual. Ñamku ta ütrüfnagomi kiñe pichuñ fey ta tuy tachi weche *Mankian*, feyta ngenlafkengepuy *Dumpall* pingetuy. Tachi ngen lafken konün antu püle, kiñeuküley tachi Pillan engu willi kürüf püledegüñ tuwülu, ka feyti *Anchümallen* tripan antu tuwülu, ka tachi *Witranalwe*, pikun kürüf püle tuwülu futra ülmen kupal.

RAPTORS

Yahgan: unknown

Mapudungun: **Ñamku,** Ñancu, Nangu

Spanish: **Aguilucho,** Aguilucho común

English: Red-Backed Hawk

Scientific: *Buteo polyosoma*
                (Accipitridae)

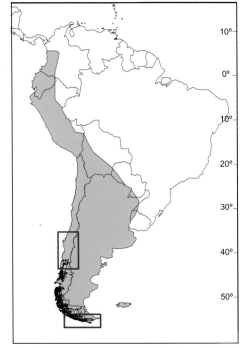

Mapuche Territory ☐

Yahgan Territory   ☐

Year-round Resident ⬤

In a variety of forest types
Sighting probability: 3

IDENTIFICATION	17.3-21.6" (44 – 55 cm) Medium-sized raptor, with a contrasting coloration of white underparts, and dark brown reddish or grayish upperparts. The white tail, with a black sub-terminal band, is often a diagnostic field mark. Yellow legs, iris brown, and dark bill. It has a wide variety of *morphos* or plumages. Adult has upperparts that are dark gray, while underparts are white. Wing tips do not extend beyond the tip of the tail.
HABITAT	Inhabits from the coast to the high Andes, up to 5,000 m, in a variety of forest types, shrublands, and open habitats, including the Patagonian steppes. It can be found from the coast to mountain valleys. It includes two subspecies: *Buteo polyosoma polyosoma* from Colombia to Cape Horn, and *B. p. exsul* endemic to Juan Fernandez archipelago.
HABITS	Alone, or in pairs, and occasionally in family groups. It glides at considerable heights, and uses a variety of perches, such as cliffs, rocks, trees or posts, which offer a good panoramic view. It emits a high pitched call while flying, or protecting its nest.
DIET	Feeds on small vertebrates, mostly mammals, and less frequently birds, reptiles and, insects.
CONSERVATION	The most common *Buteo* in Chile, and Andean South America, but it includes a subspecies restricted to Juan Fernandez Archipelago, which deserves special attention. It is classified as a species of Least Concern (IUCN).

RAPTORS

Kulalapij
Kokoriñ
Peuco
Bay-Winged Hawk

CD 1 / Track 48

In the rainforests of the Mapuche territory in the Andes Cordillera that rises above the Malleco River, the name of the village of Melipeuco denotes the place of encounter of four (*meli*) Bay-Winged Hawks (*peuco*). This raptor also receives the *Mapudungun* name of *kokoriñ* and is considered a bird of astute and sharp perception. It has strength, good eyesight and even better hearing. In the area of Chillán, when the mothers do not like the suitors of their daughters, they refer to him as the *peuco*. It is common to hear them say, "I don't like this *peuco* one little bit." On the other hand, the men refer to the "*peucas*" as the girls of ill repute.

The Bay-Winged Hawk or Harris's Hawk is a large dark brown bird of prey, whose territory extends from the southern United States to southern Chile and Argentina. It is quite frequent down to the area of Aysen and Calafate, but occasionally is found further south.* It usually perches among the branches of trees or bushes, as well as on fence posts in the country. It preys upon hares, rabbits, rodents, reptiles and many types of birds, including barnyard fowl—which leads many people in the countryside of southern Chile to call the Bay-Winged Hawk "*come pollo*," or chicken-eater. In the countryside, it is frequently hunted and persecuted, because it is said: "*kokoriñ rume iñfitukey achawall mew* ('the Harris's Hawk is damaging because it eats the chicken')." Underscoring its reputation for capturing hens and chicks in the farmyards, the poet Aillapan calls this hawk *weñefe üñüm*, or chicken thief-bird. Likewise, in the valley of the Ñuble River, the children have a game that mimics the hawk's mannerisms. The boys and girls act like the hawk and hen, chasing each other. As they play, they ask each other questions, such as "*peuco*, where do you come from," "from the rushes bog," "why do you come," "to hunt your chicks," "hunt them, if you can!"

* The Yahgan grandmothers Úrsula and Cristina Calderón recognize the Harris's Hawk in pictures, and ornithologist Steve McGehee observed one individual of Parabuteo unicinctus *on Navarino Island. This represents a new record, which considerably extends the latitudinal range known for this species.*

RAPTORS

In spite of the stigmas it carries for being the "chicken-hunter," the Bay-Winged Hawk, like other birds of prey, helps farmers and livestock owners by contributing to the control of rodent, bird and reptile populations in the countryside, shrublands, and austral forests.

Yahgan: **Kulalapij,** Kilakimöš, Cazapollo

Mapudungun: **Kokoriñ,** Peuco, Peucu, Kokori

Spanish: **Peuco,** Gavilán mixto

English: Bay-Winged Hawk, Harris's Hawk

Scientific: *Parabuteo unicinctus* (Accipitridae)

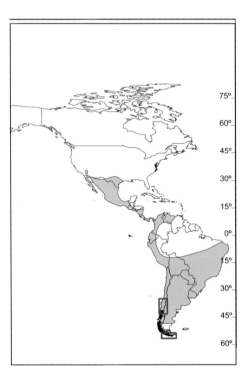

Mapuche Territory ☐

Yahgan Territory ☐

Year-round Resident ⬤

In a variety of forest types, shrublands and open habitats
Sighting probability: 2

IDENTIFICATION	18 - 22.4" (46 – 57 cm) Medium sized, slim and long-tailed raptor with general, uniform dark brownish plumage, except for the white undertail, and the tip of its tail. Strong, bluish-black beak, with yellow cere. Its legs are also yellow. Juveniles have a whitish throat, pale underparts, and a blackish tail.
HABITAT	It inhabits forest margins, shrublands, open fields such as grass, mountainous hillsides, and also urban habitats, from the coast to the Andean pre-cordillera (2,000 m). *Parabuteo unicinctus* is found from southern North America to southern South America. It includes several subspecies, and only P. u. *unicinctus* inhabits South America; in Chile extends from Arica (18°S) to Aysen (45°S), and in Argentina it reaches southward to Neuquen.
HABITS	Generally solitary, and rather sedentary. Its flights are characterized by long and high glides, but it also has fast flights to hunt its prey. It is shy and silent, but emits a loud, harsh scream.
DIET	Feeds on small to medium-sized vertebrates, occasionally including domestic birds and city pigeons, as well as insects.
CONSERVATION	It has a very wide distribution, and the subspecies P. u. *unicinctus* is more common in its southern range. It is classified as a species of Least Concern (IUCN).

Weziyau

Mañke

Cóndor

Andean Condor

 **CD 1** / Track 49

For the Mapuche, and for many other cultures, the Andean Condor, or *mañke*, is the king of birds. With a wingspan of more than 3 meters it is the largest flying bird in the Americas. For the Mapuche, *mañke* also symbolizes the Andean Cordillera with its awesome size, its snow-white ring of neck feathers, and its black body, which suggests the Cordillera's rocks and minerals. The king of birds flies at high altitudes and embodies the fundamental virtues of the Mapuche culture. The young man *Mañkian* derives his name from the expression *mañkean*, which means a man who wants to be like the Andean Condor (see the story of the Red-Backed Hawk). The surname *Mankelef* denotes a swift person, or condor, that maintains justice, who unites in himself with the attributes of: i) *Kimche* or a wise person (*kim* = wisdom, *che* = person), ii) *Norche* or a person that loves justice (*nor* = rights), iii) *Kümeche* or a good-natured person (*küme* = good) and iv) *Newenche* or a powerful person, community ruler (*newen* = spiritual or physical strength and riches, like that of Caupolicán's valiant *longko* or Mapuche chief who gave his life for the freedom of his people).

RAPTORS

The Andean Condor achieves altitudes that exceed 7,000 meters, when flying over the Andean Cordillera. The ancestral stories of the high elevation Andean cultures tell how *Viracocha* emerged from Lake Titicaca. *Viracocha* then created the sun with his light, the rain with his tears and the sky, the stars, humans, and other living beings that inhabit the region. With its majestic grandness, the Andean Condor or *cuntur* is *Viracocha's* guardian bird. The figure of *Viracocha*, sculpted in stone 2,200 years ago, remains silent today, looking towards the sunset in the Sungate at Tiahuanaco, Bolivia. This representation of *Viracocha* is surrounded by 48 winged figures: 32 have human faces and 16 possess those of the condor. This figure reminds us how linked the natures of humans and birds are. Many Andean peoples believe to be descended from the *cuntur*. Like the Mapuche, their cosmology maintains that humans are connected by the Andean Condor to the high, alpine ecosystems with their rains, rivers and lakes, the heavens with the sun, moon, stars, soils, and the waters with all the beings that inhabit them.

The modern nations of South America also admire the Andean Condor; Colombia, Ecuador, Bolivia, and Chile, have made it their "national bird." On the national shield of Chile, the characteristic crest of the male condor is represented, tellingly, as a king's crown. Until the XVIII century, this king of birds flew over the whole length of the Andean Cordillera from the Sierra de Santa Marta (Colombia) to Cape Horn. Throughout this range, the Andean Condor abounded from the peaks to sea level. As the poet Pablo Neruda writes, they are birds "born between the mountains and the marine spray." The tallest mountain in the capital of Chile is called Manquehue, which means "place of the condors" (*mañke* = condor; *hue* = place). This indicates that the Andean Condor, *cuntur* in Quechua, or *mañke*, in *Mapudungun*, was abundant in the lands that, today, are occupied by large cities in the Andean region of South America.

Paradoxically, the king of birds, so admired by diverse cultures, has disappeared from many parts of its range. In the city of Mérida, Venezuela it was seen for the last time in 1912. In Quito, Ecuador it is necessary to travel many hours on expeditions from the city to see the few condors that take refuge on the sides of Andean volcanoes. In Chile, *Vultur gryphus* is considered a species with serious conservation problems. Andean Condors are particularly vulnerable species because they occupy large territories. They can travel up to 150 kilometers in a day to find carrion. They are also very long-lived organisms—living more than 50 years—and begin to breed at the age of six. They breed every two years, laying one or two eggs.

The risk that the Andean Condor may become extinct could also involve the disappearance of the fundamental virtues embodied by it: wisdom (*kim*), justice (*nor*), goodness (*küm*) and discipline (*newen*). Like these values that we yearn and reach for, the Andean Condor is frequently difficult to see because it often soars so high that it is not detectable by the naked eye. Fortunately, in the coastal forests of the extreme southern archipelagos of South America, the Andean Condor is still clearly observable while it descends to feed on littoral ecosystems. So, in these remote forests, *mañke* or *waliáo*, astonishes us with its beauty and inspires a love for birds and for the highest values humans can cultivate.

RAPTORS

MAPUCHE STORY

 CD 1/ Track 67

The Condor or the *mañke* is for the Mapuche culture the King of the Birds. It is a symbol of the mountains, not only due to its great size and impressive wing-length, but also its black and white colors, which signify the snow-capped peaks and the black mountain rocks. For the Mapuche, the condor unites the virtues of being *Kimche*, a wise person, *Norche*, a person who loves justice, *Kümeche*, a good-natured person, and *Newenche*, a powerful, lordly person. This King of Birds is the national bird of Columbia, Ecuador, Bolivia and Chile. *Manke* was once abundant on *Manquehue* Mountain, which means place (*hue*) of the condors (*mañke*), in the center of Santiago, the capital of Chile. It is still found today throughout the territory of the austral forests, but paradoxically, this king of birds, this national symbol of four countries, finds itself now threatened with extinction.

MAPUDUNGUN VERSION

Affkappe Affkappe Affkappe Affkappe
Affkappe Affkappe Affkappe Affkappe
Tachi mañke fey ngennidol puwerra üñümtuley pu mapuche tañi kimün meu. Pirentuku winkul püle rekeley. Fey tañi fütramongen ka feyñi lügkürü femngen. Feyta chi mañke kom fütra kim newen dungu wepümniyey kimche inanielu wera kimün, Norche nentunielu norum dungu, Kümeche llakonagduam, Newenche feyta ülmen ka fütra ñidol. Welulelureke, ngen ñidol puwera üñüm mülelu willi mapu püle cheuñi mülen mawidantu ka rumel pirentuku winkul tripan antüpüle komtrokiñ willi admapupüle, fey ñidol üñüm pingey Colombia, Ecuador, Bolivia ka Chülle. Mapu (Chile), kafey rume mülekefuy winkul Mankehue ñidol waria ta Chile, feula fey epeke ngewetulay larumelelu.

Yahgan:	Weziyau, Wañiáo
Mapudungun:	**Mañke**, Manque
Spanish:	**Cóndor**, Cóndor andino
English:	Andean Condor
Scientific:	*Vultur gryphus* (Cathartidae)

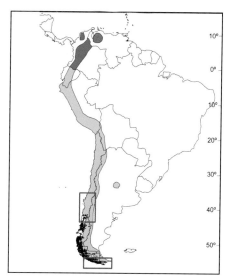

Mapuche Territory ▢

Yahgan Territory ▢

Year-round Resident ●

Occasional Visitor ●

In a variety of forest types, mountainous and coastal open habitats
Sighting probability: 2

IDENTIFICATION	37.4-48" (95-122 cm) Majestic glider, it is the heaviest flying bird in the Americas, reaching up to 11 to 15 kg (24-33 lb) for males and 6 to 14 kg (13-31 lb) for females. Its wingspan ranges from 280 to 320 cm (110-126"). In flight, the long, rectangular shaped wings exhibit on their edges prominent primary feathers, which are visible from long distance. This large black vulture has a distinctive ruff of white feathers surrounding the base of its neck, and large white patches on the wings, especially in the male. The male exhibits a conspicuous dark gray or red comb, and both sexes have featherless, reddish head and neck. It has a strong, hooked, bone-color beak, and the irises are light brown in males, and red in females. Contrary to the general rule among birds of prey, the female is smaller than the male.
HABITAT	The Andean Condor is found along the Andean cordillera from Venezuela and Colombia to Cape Horn. It soars over open grasslands and alpine zones up to 5,000 m (16,000 ft), over lowland desert areas in Chile and Peru, and the Pacific coasts of western South America, especially in the sub-Antarctic ecoregion.
HABITS	Alone, in pairs or in groups, occasionally forming large flocks when feeding on a carcass. In flight, after attaining a moderate elevation, it rarely flaps its wings, relying on thermals to stay aloft. It nests in caves or cavities in high vertical cliffs in the mountains, and on coastal cliffs in the sub-Antarctic ecoregion. Andean Condors reach sexual maturity when the bird is five or six years of age, and it lays one or two eggs every other year.
DIET	The condor is primarily a scavenger, feeding on carrion, preferring large mammals' carcasses, such as those of guanacos or cattle. Also robs bird nests or preys on new born lambs, and occasionally it attacks sick or weakened cattle.
CONSERVATION	It has a moderately small global population, which is suspected to be declining due to habitat degradation and persecution by humans. It is classified as Near Threatened by IUCN. Today, it is very rare in Venezuela and Colombia, and its main populations are in Argentina, and Chile. The Chilean Hunt Law and Regulations consider this species as "beneficial for the equilibrium of ecosystems," also "beneficial for agriculture, forestry, and husbandry activity," and "Vulnerable" in most of its territory (except the extreme south). Therefore, its hunting is strictly forbidden. The Andean Condor also receives legal protection under several international treatises and conventions, including CITES (Convention on International Trade in Endangered Species of Wild Flora and Fauna). **NT**

RAPTORS

Iloéa
Kelüwún Kaniñ
Jote de cabeza colorada
Turkey Vulture

(•) **CD 1** / Track 50

The Turkey Vulture is found in almost all the Americas and in a great variety of environments. From close up, it is distinguishable by its red legs, head, and neck. Its *Mapudungun* names *kelwi* or *kelüwún kaniñ*, allude to its red color (*kelü* = red, *Kaniñ* = vulture, *wün* = mouth and *wünüñüm* = beak); which is to say bird (*üñüm*) mouth (*wün*).

Recently, it was discovered that these vultures have a very developed sense of smell that permits them to discern and locate decomposing meat from great distances. In its flights over temperate and tropical forests, these carrion eaters have the ability to detect food that is on the forest floor below the canopy.

When they nest, Turkey Vultures prefer to use caves, but in southern Chile and Argentina they often use low brushland or fallen trunks in the forests, and occupy the same nest for several years. In the Magellanic Archipelago Region, where the evergreen forests grow down to the high tide line, Turkey Vultures are abundant in the littoral zone. In this forest-shore ecotone they nest in dense vegetation, and eat a great variety of carrion, especially in colonies of sea lions and penguins.

Vultures provide a great service to ecological systems by cleaning the bones of dead animals, thereby contributing to the chain of decomposers that then turn dead organic matter into soil. For humans, Turkey Vultures provide a special service by eliminating trash, and removing rotten carcasses. *Lafkenche* poet Lorenzo Aillapan illustrates this ecological-cleaning service of vultures, writing in the *Twenty Winged Poems*:

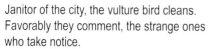

Janitor of the city, the vulture bird cleans.
Favorably they comment, the strange ones who take notice.
Full-time cleaner of the beach and the open countryside,
The people of the place comment with admiration.

Tachi kaniñ üñüm liftu wariafe
Pikeeyu ka pu che inarumengelu
Liftu küyüm rüpüfe ka feyti lelfünche
Feypilekey pu che lofmeu

RAPTORS

RAPTORS

Yahgan:	**Iloéa,** Īliūaia, Aiskalaix, Akixlaiix, Ilaóea
Mapudungun:	**Kelüwún Kaniñ,** Queluy, Quelvoni, Kanin, Kelwi
Spanish:	**Jote de cabeza colorada,** Gallinazo, Jote de cabeza roja
English:	Turkey Vulture, Turkey Buzzard
Scientific:	*Cathartes aura* (Cathartidae)

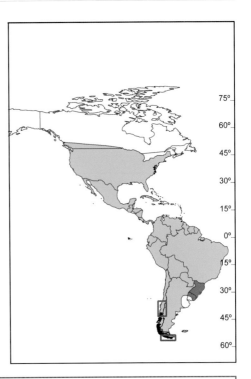

Mapuche Territory ☐

Yahgan Territory ☐

Year-round Resident ●

Occasional Visitor ●

In a variety of forest types and open habitats
Sighting probability: 3

IDENTIFICATION	22-32" (66-81 cm) Large bird (wingspan = 68–72"; 173–183 cm) with black plumage, and a featherless, purplish red head and neck. In flight, it is distinguishable from other vultures by the silver-gray color of its underwing flight feathers, which contrast with the black of the narrower, underwing lining covers. It has a strong, short, hooked, ivory-colored beak, and its legs and feet are pink-skinned. There is minimal sexual dimorphism, although females tend to be slightly larger.
HABITAT	The Turkey Vulture is the most abundant vulture in the Americas, inhabiting from southern Canada to Cape Horn (56°S). It includes five subspecies, of which only *C. a. jota* (the Chilean Turkey Vulture) is present in the South American temperate forest biome. *C. a. jota* inhabits a variety of open and semi-open habitats, including forests, shrublands, pastures, wetlands, deserts, as well as farmlands and cities, from the coast to the Andean pre-cordillera (2,000 m). It is found in Argentina from Jujuy to Tierra del Fuego, Staaten Islands and Malvinas Islands, and in Chile from Arica to Cape Horn. In the sub-Antarctic ecoregion it is especially frequent in marine mammal and bird breeding colonies.
HABITS	Alone, in pairs, or in flocks, especially around a dead animal. As all vultures, it is an excellent glider. It soars at low altitude because it finds its food using both its vision and sense of smell, sensitive to gasses produced by decay processes of recently deceased animals.
DIET	Scavenger, it feeds on a variety of carrion, and it reaches maximum densities along the sub-Antarctic archipelago where it feeds on carrion, and also newborn sea lions and other mammals and birds.
CONSERVATION	It has a very large range, and it is classified as a species of Least Concern (IUCN). However, *Cathartes aura* is strictly protected by the Chilean Hunt Law and Regulations, which forbids its hunting because, this bird is "beneficial for agriculture, forestry, and husbandry activity." This species also receives legal protection under several international treatises and conventions for migratory birds.

Kaniñ kürüwún
Jote de cabeza negra
Black Vulture

 **CD 1** / Track 51

The *Mapudungun* names of the vultures describe the color of the head of the two species that inhabit the southern forests of South America. As such, the Red-Headed (or Turkey) Vulture is *Kaniñ kelülonko* (*Kaniñ* = vulture; *kelü* = red; *lonko* = head) and the Black Vulture is *Kaniñ kürülonko* (*kürü* = black).

The Black Vulture inhabits both the American continents, but does not reach the high latitudes that the Turkey Vulture does. In Chile and Argentina, they reach a maximum southern range at 50ºS, while the Turkey Vulture reaches Cape Horn (56º S).

Black Vultures are social birds and are frequently seen soaring in concentric circles, where groups of individuals cooperate in the search for a dead animal, using sense of sight in open habitats. The *Lafkenche* Bird-Man, or *Uñumche*, Lorenzo Aillapan poetically celebrates the social behavior of the vultures, singing:

With the rythmn of their animated flight, they demonstrate that they are party-goers of renown. Performing their favorite birdly job with an air of solidarity, circling about their preferred land, with calculation. *¡Affkappe affkappe affkappe affkappe arolchi müleputuan gañi aliwen meu!*	*Fey mütamütem fentren üpün amuleyngün Rüf feymeu kollefe pingekey üytungen Feychi kudau rumel nielukay komengün* *Fill wall lelfün mapu müpüg adkintuyawingün. ¡Affkappe affkappe affkappe affkappe arolchi müleputuan gañi aliwen meu!*

In the margins of the austral forests it is common to observe groups of Black Vultures perched on trees, where they usually extend their wings to the sun to warm their body and to dry and align their feathers. Although vultures prefer recently dead animals, they eat all types of carrion; hence, vultures are seen as good cleaners. For this reason, when the *Williche* on Chiloe Island find an egg of a *Kaniñ kürüwún* among their crops, they foresee in this a message of good luck, and they predict good harvests.

RAPTORS

Yahgan: out of bird's range

Mapuche: **Kaniñ kürüwún**, Queluy, Quelvoni

Spanish: **Jote de cabeza negra**

English: Black Vulture

Scientific: *Coragyps atratus* (Cathartidae)

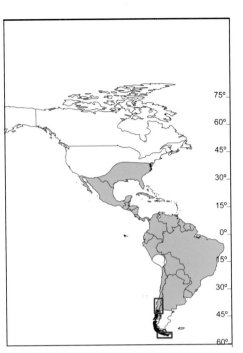

Mapuche Territory ☐

Yahgan Territory ☐

Year-round Resident ⬤

In a variety of forest types and open habitats
Sighting probability: 4

IDENTIFICATION	21.5-29.5" (55-75 cm) Large bird (wingspan = 53–63"; 133–160 cm) with black plumage, and a featherless, dark gray, wrinkled-skin head and neck. In flight, the bases of its six external primary feathers are white, producing a white patch on the underside of the wing's edge, which contrasts with the black color of the rest of the underwing. It has a longer and slender, hooked, blackish beak with a pale tip. Its legs are grayish white.
HABITAT	*Coragyps atratus* inhabits from southeastern United States of America to southern Chile. It includes three subspecies, of which only one, *C. a. foetens* (the Andean Black Vulture), is present in the South American temperate forest biome. *C. a. foetens* inhabits a variety of open and semi-open habitats, but prefers coastal and lowland forests, shrublands, pastures, wetlands, and deserts; it ventures into the Andean pre-cordillera (up to 2,000 m), and has a predilection for dumps, farmlands and cities. It is found in Argentina from Jujuy to Chubut, and in Chile from Arica to Aysen (45°S).
HABITS	Gregarious scavenger, occasionally forming very large flocks around carrion. It perches in a distinctive upright posture on trees, posts or buildings. Walks or jumps over the ground. It soars at high altitudes, holding its wings horizontally, and interpersing vigorous and short flaps. Pairs are formed following a courtship ritual performed on the ground, where several males circle around a female with their wings partially open as they move their heads up and down. To nest they use caves and rocky cliffs. Its only vocalizations are grunts or low hisses.
DIET	It feeds on carrion, but it also eats eggs and occasionally preys on newborn animals.
CONSERVATION	It is classified as a species of Least Concern (IUCN). However, it is strictly protected by the Chilean Hunt Law and Regulations, which forbids its hunt, because it considers this species to be "beneficial for agriculture, forestry, and husbandry activity." It also receives special legal protection under international treatises and conventions for migratory birds.

THE VOICES AND STORIES MUST CONTINUE

For the grandparents and grandchildren
of the birds, *pij*, *üñüm* or *aves*
and the *che*, human beings, *yaganes* or *seres humanos*
that fly (or would like to fly)
as *omora*, *pinda*, hummingbird, *picaflor* or *Sephanoides sephaniodes*,
from flower to flower in the forests of austral South America,
feeding ourselves and feeding many other living beings,
captivated by the beauty, mystery and delicacy
of the odors, tastes, colors, textures and sounds
that emanate from this inexhaustible fount of biological and cultural diversity,
flying over the indifference and the dogmatism,
determined to live the risk and vitality of sharing the life with whom we co-inhabit
these rainy and fecund *forests, ashuna* or *mawida* of the planet,
we must continue listening and permitting the expression of
their multiple voices.

ACKNOWLEDGMENTS

Acknowledgments for the first edition of this book in 2003:

As it was said in the Introduction, this guide is a collective work that is possible thanks to the fact that birds, forested ecosystems and very diverse people exist. In particular, the authors sincerely thank the following individuals and organizations:

Mario Chiguay, President of the Indigenous Yahgan Community of Bahía Mejillones, and all the other members of the community who collaborated in this project, especially Julia González;

Manuel Muñoz, Advisor to the Chiloé Council of Chiefs, the members of the Huilliche communities in Chanquin and Huentemó, the school teachers of Chanquin and Mr. Francisco Delgado, Chiloé National Park, who encourage and helped the initiation of this project;

The poet Lorenzo Aillapan, who teaches in the Pullümapukimunweftuy Mapuche Academy in Puerto Saavedra, a cultural center which this guide hopes to serve;

Dr. Víctor Fajardo, President of the University of Magallanes, Luis Oval, Vice-President for Academic Affairs, and the professors of the Department of Natural Science Orlando Dollenz, Sylvia Oyarzún and Dr. Andrés Mansilla; this guide book also forms part of the biocultural conservation and environmental ethics program, initiated by the University of Magallanes and the Omora NGO at the Omora Ethnobotanical Park in the Chilean Antarctic Province;

The Chilean Antarctic Provincial Government and Governor Eduardo Barros, who actively participates in the search for equity and conservation at the southern tip of the Americas, and graciously received poet Lorenzo Aillapan into the provincial government's guesthouse and supported the work of this guide;

The National Natural History Museum, which provided access to its valuable collection that demonstrates Chile's biological and cultural diversity; we especially acknowledge the cooperation of the museum's director, Eliana Ramírez, and Curator of Ornithology Juan Carlos Torres-Mura, who has been key in completing this project;

The Center for Biodiversity and Conservation at the Department of Ecology and Evolutionary Biology – University of Connecticut (USA), which permitted the necessary research for the preparation of distribution maps, as well as the manuscript of this book;

Ricardo Rozzi profoundly thanks the members of his doctoral committee John Silander, Gregory Anderson, Robin Chazdon, Robert Colwell and Steward Pickett, as well as Robert Dewar, and Christopher Elphick, at the University of Connecticut for their ecological, anthropological and ornithological guidance during the development of this guide book;

The cyber and personal exchanges by way of the Network of *Enseñanza de la Ecología en el Patio Escolar* (*EEPE*), which remains a fount of learning and constant stimulation; particular thanks to

213

the network's coordinators Peter Feinsinger, Alejandro Grajal, and Ricardo Stanoss of the Audubon Society's Latin America and Caribbean Program (USA), we hope that this guide will be an educational source material to aide in the search for ways to live respectfully with the world's biological and cultural diversity by showing the urban, rural and remote corners of Latin America;

Leonard Yannielli and the volunteers from the Earthwatch Institute project *The Owls of Cape Horn* for their collaboration in the recording of birdcalls and the study of the natural history of the avifauna in the austral extreme of the Americas;

The Senda Darwin Foundation, especially its directors Juan Armesto and Mary Willson, and Claudia Hernández, Claudia Papic, Marco Méndez, Emma Elgueta and Iván Díaz, who cooperated in the ornithological research; Hernán Rivera from the Chilean national parks service and Guillermo Egli, of the Union of Chilean Ornithologists, with whom we organized the workshop *Recognizing the Forest Birds of Chiloé*, conducted in Chiloé National Park in December 1995. In this workshop participated: Francisco Delgado, José Gallegos, Mario Guinao, Jeno Muñoz, Jorge Panichini, Roberto Rosas, Jose Subiabre and Carlos Uribe (Chilean national parks agency), and Eduardo Ramilo, Juan Salguero and Lorenzo Simpson (Argentine parks delegation), and Enrique Couve, Luis Espinoza and Andreas von Meyer (Union of Chilean Ornithologists), and Carlos Sabag (Senda Darwin Foundation), and Andrea Bachmann and Lily Sheeline (Peace Corps – USA); all helped precipitate the initiation of this book;

The recording studios of Button Records (Puerto Williams) and Radio WHUS-University of Connecticut (USA), which offered their facilities and studio time, under the direction of Nolberto González and John Schwenk, respectively, for the editing of sound tracks that allowed the combination of bird calls with stories and the voices of different people in the CDs of this guide; without this technical and infrastructure cooperation for the sound archives, this guide would not have been possible;

The composers and students of Taller Matta 365, Valparaíso, who have given new life to the multiple bird and human voices in their experimental music workshop; Andrés Alcalde for his music and for the piece for piccolo *Gli Uccellini*, played by Guillermo Lavado;

Patricio Riquelme for his friendship, his art and the design of the CD cover; artists Daniel Martínez and Jay Barry for their drawings (see Photograph and Figure Credits) and all those who donated their photographs (see Photograph and Figure Credits);

Enrique Couve and Claudio Vidal, Fantástico Sur, for welcoming us in the alliance between science, educational tourism and birds;

Humberto Maturana, Baird Callicot and Eugene Hargrove for their works on experimental epistemology, comparative environmental ethics and the intrinsic value of all living beings that inspire this guide;

Mary Kalin-Arroyo, and Juan Armesto, director and vice-director of the Institute of Ecology and Biodiversity Research - University of Chile, have provided a continuous intellectual and institutional support to develop transdisciplinary research and to integrate research and education; Jorge Pavez, Fondo de las Americas, encouraged in 1999 the development of the project *Biological and Cultural Conservation at the Southern End of the Americas* that permitted the establishment of biocultural conservation and environmental ethics program initiated with the University of Magallanes;

The members of the Omora Foundation, especially Héctor Massardo, Andrés Marin, Maurice van de Maele, Livia Firmani, Bryan Connolly, Augustin Berghöfer, Margaret Sherriffs, Alicia Lavanchy and Alejandro Correa, and to all the persons who made this educational project possible from 2000 to 2003.

Acknowledgments for this new 2010 edition

For this revised edition by University of North Texas Press & Ediciones Universidad de Magallanes we especially thank the valuable suggestions and comments made by the reviewers Christopher Elphick, ornithologist from the Department of Ecology and Evolutionary Biology of the University of Connecticut, and David Rothenberg, philosopher and musician from the Department of Humanities of the New Jersey Institute of Technology; and the continuous support of Gene Hargrove, Director of the Center for Environmental Philosophy of the University of North Texas, who encouraged us for the publication of this second edition;

The work team for this revised edition was led by Chilean artist Paola Vezzani who worked with the graphic designer Alejandra Calcutta, Kelli Moses, and Francisca Massardo in the design and layout of the book, and on the maps with María Rosa Gallardo, engineer, Universidad de Magallanes; the translation from Spanish to English were done and/or revised by Christopher Anderson, Kurt Heidinger, and Ricardo Rozzi in the first edition of this book, and were further revised and edited with the valuable help of Kelli Moses for this revised edition; José Cristóbal Pizarro helped with the preparation of the ornithological tables; we sincerely appreciate the academic support by Irene Klaver, and J. Baird Callicott;

Dr. Claudio Wernli, Director of the Millennium Science Initiative Program (MSI) that provides support for the Institute of Ecology and Biodiversity, Mr. Bernardo Troncoso, Director of the CORFO-Chile (*Corporación de Fomento de la Producción*, Corporation for the Foment of Production), Magallanes and the Chilean Antarctic Region, CORFO office for the Magallanes and the Chilean Antarctic Region, and the INNOVA-CORFO project 08CTU01-22 2009-2011 Ecotourism with a Hand-Lens, Mr. Sergio Lausic, director of the Editorial Board of the *Ediciones Universidad de Magallanes*, Dr. Juan Oyarzo and Dr. José Maripani, Vice-Presidents, Universidad de Magallanes, Dr. Earl Gibbons, Vice Provost and Associate Vice President for International Education, Mr. Ronald Chrisman, Director of the UNT Press, and Ms. Karen DeVinney, Managing Editor from the University of North Texas Press, made it possible to commence the international series of publication based on the collaboration of *UNT Press & Ediciones Universidad de Magallanes*, in association with the Sub-Antarctic Biocultural Conservation Program, which is initiated with this *Multi-Ethnic Bird Guide of the Sub-Antarctic Forests of South America*.

SELECTED BIBLIOGRAPHY

References concerning ethnography and biocultural conservation in the South American temperate forests

Agusta, F. F. J. d. 1903. *Gramática Mapuche Bilingüe*. Ediciones Séneca, Santiago, Chile.

Aillapan, L. & R. Rozzi. 2004. Una etno-ornitología mapuche contemporánea: poemas alados de los bosques nativos de Chile. *Ornitología Neotropical* 15: 419-434.

Arnold, J. 1996. The inverse system in Mapudungun and other languages. *Revista de Lingüística Teórica y Aplicada* 34.

Bart, F. 1948. Cultural development in Southern South America: Yahgan and Alakaluf vs. Ona and Tehuelche. *Acta Americana* 6: 192-9.

Bate, L. 1983. Comunidades primitivas de cazadores recolectores en Sudamérica. *Historia General de América*, Volumen 2. Caracas, Venezuela.

Bird, J. 1946. The Archaeology of Patagonia. En *Handbook of South American Indians*, Smithsonian Institution, Washington, USA.

Bridges, E.L. 1949. *Uttermost Part of the Earth*. E.P. Dutton and Company Inc., N.Y. (Special ed. 1950).

Bridges, T. 1933. *Yamana-English Dictionary*. F. Hestermann & M. Gusinde (Eds.). Second Edition (1987) Zagier y Urruty Publicaciones, Buenos Aires.

Cárdenas, R. 1994. *Diccionario de la lengua y de la Cultura de Chiloé*. Olimpho, Santiago.

Catrileo, M. 1988. *Mapudunguyu: Curso de Lengua Mapuche*. Universidad Austral de Chile.

Catrileo, M. 1998. *Diccionario Linguístico-Etnográfico de la Lengua Mapuche. Mapudungun-Español- English*. Editorial Andrés Bello, Santiago de Chile. 3°ed.

Chapman, A. 2006. *Martin Gusinde: Lom, Amor y Venganza. Mitos de los Yamana de Tierra del Fuego*. LOM Ediciones, Santiago, Chile.

Chapman, A. 1986. *Los Selk'nam. La vida de los Ona*. Emecé Editores, Buenos Aires.

Charno, J., L. Blumenfeld, D. Lewiston & B. Wentz. 1993. *Voices of Forgotten Worlds. Traditional Music of Indigenous People*. Ellipsis Arts, N.Y.

Citarella, L. (ed.). 1995. *Medicinas y Culturas en la Araucanía*. Editorial Sudamericana, Santiago, Chile.

Coña, P. & Ernesto Wilhelm de Moesbach (1930). *Vida y Costumbres de los Indígenas Araucanos en la Segunda Mitad del Siglo XIX* (Pascual Coña & Ernesto Wilhelm de Moesbach). Imprenta Cervantes, Santiago, Chile.

Cooper, J. 1946a. The Ona. En *Handbook of South American Indians*. Volume 1. Smithsonian Institution, Washington D.C., pp. 107-125.

Cooper, J.M. 1946. The Yahgan. En *Handbook of South American Indians. The Marginal Tribes*. Vol 1. J. H. Steward ed. Smithsonian Institution, Bull. 143. U.S.Government Printing Office, Washington. pp. 81-106.

Croese, R. (1980). *Estudio Dialectológico del Mapuche*. Universidad Austral de Chile, Valdivia.

Dannemann, M., & A. Valencia. 1989. *Grupos Aborígenes Chilenos. Su Situación Actual y Distribución Territorial*. Universidad de Santiago de Chile, Instituto de Investigaciones del Patrimonio Territorial de Chile, Colección Terra Nostra N° 15.

Darwin, C. 1838. *The Voyage of the Beagle*. Reprint, London: Everyman's Library, 1975.

Darwin, C. 1871. *The Descent of Man*. Princeton University Press, ed. (1981). Princeton, New Jersey.

Emperaire, J., 1963. *Los Nómades del Mar*. Ediciones de la Universidad de Chile, Santiago de Chile.

Erize, E. 1960. *Diccionario Comentado Mapuche-Español*. Cuadernos del Sur, Universidad Nacional del Sur, Buenos Aires, Argentina.

Gorgoglione, E.C. 1997. Guía de Campo para la Identificación de las Aves del Neuquen. 1°ed. Editorial Hemisferio Sur, Montevideo, Uruguay.

Grimes, B. (ed.). 2000. ETHNOLOGUE: Languages of the World. 14th Edition, Summer Institute of Linguistics, Dallas, Texas.

Gusinde, M. 1946. *Urmenschen im Feuerland*. Paul Zsolnay Verlag, Berlin.

Gusinde, M. 1961. *The Yamana: The Life and Thought of the Water Nomads of Cape Horn*. Volumes I-V, translated by F. Schutze. New Haven Press, USA.

Gusinde, M. 1968. *Expedición a Tierra del Fuego*. Editorial Universitaria, 2° edición, Imprenta Salesianos, Santiago.

Hernández, A., N. Ramos & C. Cárcamo. 1997. *Diccionario Ilustrado Mapudungun Español Inglés*. Pehuen, Santiago, Chile.

Hidalgo, J., V. Schiappacasse, H. Niemeyer, C. Aldunate & P. Mege (eds.). 1996. *Etnografía. Sociedades Indígenas Contemporáneas y su Ideología*. Editorial Andrés Bello, Santiago, Chile.

Hyades, P & J. Deniker. 1891. *Anthropologie, Etnographie. Mission Scientifique du Cap Horn. 1882-1883*. Tome VII. Gauthier-Villard et Fils, Imprimeur-Libraries, Paris.

King, Captain. 1839. *Narrative of the Surveying Voyages of His Majesty's Ships Adventure and Beagle, Between the Years 1826 and 1836*. Henry Colburn. V 1., London.

Koppers, W. 1997. Entre los Fueguinos. Ediciones de la Universidad de Magallanes, Punta Arenas, Chile.

Latcham, R. 1928. *La Prehistoria Chilena*. Editorial Universo, Santiago de Chile.

Legoupil, D. 1993. El Archipiélago de Cabo de Hornos y la costa sur de la Isla Navarino: poblamiento y modelos económicos. *Anales del Instituto de la Patagonia* 22: 101-121.

Lenz, R. 1905-1910. *Diccionario Etimolójico de las Voces Chilenas Derivadas de las Voces Indíjenas Americanas*. Imprenta Cervantes, Santiago, Chile.

Massardo, F. & R. Rozzi. 2004. Etno-ornitología yagán y lafkenche en los bosques templados de Sudamérica austral. Ornitología Neotropical 15: 395-407.

Massardo F. & R. Rozzi. 2006. *The World's Southernmost Ethnoecology: Yahgan Craftmanship and Traditional Ecological Knowledge*. Bilingual English-Spanish edition. Ediciones Universidad de Magallanes, Punta Arenas, Chile.

McEwan, L.A. Borrero & A. Prieto (eds.). 1997. *Patagonia: Natural History, Prehistory and Ethnography at the Uttermost End of the Earth*, Princeton University Press.

Mena, F. 1988. Cazadores recolectores en el área patagónica y tierras bajas aledañas (Holoceno medio y tardío). *Revista de Arqueología Americana* 4: 134-167.

Mires, A. 2000. *Así en las Flores como en el Fuego. La Deidad Colibrí en Amerindia y el Dios Alado en la Mitología Universal*. Ediciones Abya-Yala, Quito, Ecuador.

Nichols, J. 1990. Linguistic diversity and the first settlement of the New World. *Language* 66:475-521.

Ortíz-Troncoso, O. 1996. Los últimos canoeros. En Hidalgo J., V. Schiappacase, H. Niemeyer, C. Aldunate and P. Mege (eds.). *Etnografía. Sociedades Indígenas Contemporáneas y su Ideología. Culturas de Chile*. Editorial Andrés Bello, Santiago de Chile. pp.135-147.

Piana, E.L., A. Vila, L. Orquera, & J. Estevez. 1992. Chronicles of Onashaga: Archaeology in the Beagle Channel. *Antiquity* 66:771-783

Plath, O. 1996. *Lenguaje de los Pájaros Chilenos*. (2°). 1ª edición en Editorial Grijalbo. Santiago de Chile.

Ramírez-Sánchez, C. 1989. *Voces Mapuches*. Valdivia, Chile.

Rivas, P., C. Ocampo & E. Aspillaga. 1999. Poblamiento temprano de los Canales Patagónicos: el núcleo septentrional. *Anales del Instituto de la Patagonia* 27: 221-230.

217

Roquera, L. & E. Piana. 1999. *La Vida Social y Material de los Yamana*. EUDEBA, Buenos Aires, Argentina.

Rozzi, R., J. Silander, J.J. Armesto, P. Feinsinger & F. Massardo. 2000. Three levels of integrating ecology with the conservation of South American temperate forests: The initiative of the Institute of Ecological Research Chiloé, Chile. *Biodiversity and Conservation* 9: 1199-1217.

Rozzi, R. 2004. Implicaciones éticas de narrativas yaganes y mapuches sobre las aves de los bosques templados de Sudamérica austral. Ornitología Neotropical 15: 435-444.

Rozzi, R. & A. Poole. 2008. *Biocultural and Linguistic Diversity* . In "Encyclopedia of Environmental Ethics and Philosophy," Eds. B. Callicott & R. Frodeman, Volume 1: pp.100-104. MacMillan Reference Book Gale, Cengage Learning, Farmington Hills, Michigan.

Salas, A. 1984. *Textos Orales en Mapuche o Araucano del Centro-Sur de Chile*. Editorial de la Universidad de Concepción, Concepción, Chile.

Salas, A. 1992. Lingüística Mapuche. Guía Bibliográfica. *Revista Andina* Año 10 (N°2).

Smeets, I. 1989. *A Mapuche Grammar*. Leiden, Holland.

Stambuk, P. 1986. *Rosa Yagán. El Ultimo Eslabón*. Editorial Andrés Bello, Santiago, Chile.

Villagrán, C., R. Villa, L. Hinojosa, G. Sánchez, M. Romo, A. Maldonado, L. Cavieres, C. Latorre, J. Cuevas, S. Castro, C. Papic & A. Valenzuela. 1999. Etnozoología Mapuche: un estudio preliminar. *Revista Chilena de Historia Natural* 72: 595-628.

Wilbert, J. (Ed.). 1977. Folk Literature of the Yamana Indians. Martin Gusinde's Collection of Yamana Narratives. University of California Press, London.

Wilhelm de Moesbach, P. E. 1991. *Idioma Mapuche*. Imprenta y Editorial San Francisco, Villarica, Chile.

References about ecology and conservation of South American temperate forests and their avifauna

Anderson, C. & R. Rozzi. 2000. Bird assemblages in the southernmost forests in the world: Methodological variations for determining species composition. *Anales del Instituto de la Patagonia* 28: 89-100.

Armesto, J.J, C. Villagrán & M.T. Kalin. 1996. *Ecología de los Bosques Nativos de Chile.* Editorial Universitaria, Santiago, Chile.

Armesto, J.J., R. Rozzi, C. Smith-Ramírez & M.T.K. Arroyo. 1998. Effective conservation targets in South American temperate forests. *Science* 282: 1271-1272.

Arroyo, M.T.K., M. Riveros, A. Peñaloza, L. Cavieres & A.M. Faggi. 1996. History and regional richness patterns of the cool temperate rainforest flora of Southern South America. En R.G. Lawford, P. Alaback & E.R. Fuentes (eds.), *High Latitude Rain Forest and Associated Ecosystems of the West Coast of the Americas: Climate, Hydrology, Ecology and Conservation*. Springer-Verlag, Berlin, pp. 135-151.

Axelrod, D.I., M.T.K. Arroyo & P. Raven. 1991. Historical development of temperate vegetation in the Americas. *Revista Chilena de Historia Natural* 64: 413-446.

Cofre, H. 1999. Patrones de rareza de las aves del bosque templado de Chile: implicancias para su conservación. Boletín Chileno de Ornitología 6: 8-16.

CONAF-CONAMA-BIRF. 1997. *Catastro y Evaluación de los Recursos Vegetacionales Nativos de Chile*. Universidad Austral de Chile, PUC, Universidad Católica de Temuco.

Donoso, C. 1995. *Bosques Templados de Chile y Argentina*. Editorial Universitaria, Santiago, Chile.

Erazo, S. 1987. Observaciones y alcances ecológicos a la comunidad de aves terrestres en la desembocadura del río Aconcagua, V Región, Chile. Revista Geográfica de Valparaíso (Chile) 18: 51-62.

Estades, C. & S. Temple. 1999. Deciduous-forest bird communities in a fragmented landscape dominated by exotic pine plantations. Ecological Applications 9: 573-585.

Fuentes, E. 1994. *¿Qué Futuro Tienen Nuestros Bosques?. Hacia la Gestión Sustentable del Paisaje del Centro y Sur de Chile*. Ediciones Universidad Católica de Chile, Santiago, Chile.

Gajardo, R. 1994. *La Vegetación Natural de Chile: Clasificación y Distribución Geográfica*, Editorial Universitaria, Santiago, Chile.

Glade, A (ed). 1988. *Libro Rojo de los Vertebrados Chilenos*. Corporación Nacional Forestal, Santiago (Chile).

Hinojosa, L.F. & C. Villagrán. 1997. Historia de los bosques del sur de Sudamérica, I: antecedentes paleobotánicos, geológicos y climáticos del Cono Sur de América. *Revista Chilena de Historia Natural* 70: 225-239.

Jaksic, F. & P. Feinsinger. 1991. Bird assemblages in temperate forests of North and South America: a comparison of diversity, dynamics, guild structure and resource use. Revista Chilena de Historia Natural 64:491-510.

Jaksic, F. 1997. Ecología de los vertebrados de Chile. Ediciones Universidad Católica de Chile. Santiago, Chile. 262 pp.

López, M.V. 1990. Variación estacional en el uso de los recursos alimenticios para algunos componentes de una taxocenosis de aves Passeriformes en Quebrada de la Plata, Chile Central. Tesis de Magister. Facultad de Ciencias, Universidad de Chile. 116 pp.

Muñoz, M., H. Nuñez & J. Yánez (comps.). 1996. *Libro Rojo de los Sitios Prioritarios para la Conservación de la Diversidad Biológica en Chile*, Ministerio de Agricultura-CONAF, Santiago, Chile.

Primack, R., R. Rozzi, P. Feinsinger, R. Dirzo & F. Massardo. 2001. *Fundamentos de Conservación Biológica: Perspectivas Latinoamericanas*. Fondo de Cultura Económica, México.

Rau, J., A. Gantz & G. Torres. 2000. Estudio de la forma de fragmentos boscosos sobre la riqueza de especies de aves al interior y exterior de áreas silvestres protegidas. Gestión Ambiental 6: 33-40.

Riveros G.M. & M.C. López-Calleja 1990. Distribución de las aves en el período no reproductivo y su relación con las formaciones vegetacionales presentes en el Parque Nacional la Campana, Chile Central. *Boletín Sociedad de Biología de Concepción* 61:161-166.

Rodríguez, R., O. Matthei & M. Quezada. 1983. *Flora Arbórea de Chile. Biblioteca de Recursos Renovables y no Renovables de Chile*. Editorial Universidad de Concepción, Concepción, Chile.

Rozzi, R., J.J. Armesto, A. Correa, J.C. Torres-Mura & M. Sallaberry. 1996. Avifauna de bosques primarios templados en islas deshabitadas del archipiélago de Chiloé. Revista Chilena de Historia Natural 69: 125-139.

Rozzi, R., F. Massardo, C. Anderson, A. Berghoefer, A. Mansilla, M. Mansilla, J. Plana, U. Berghoefer, E. Barros, & P. Araya. 2006. *The Cape Horn Biosphere Reserve*. Ediciones Universidad de Magallanes, Punta Arenas, Chile.

Schlatter, R. 1976. Contribución a la ornitología de la Provincia de Aysén. Boletín Ornitológico (Chile) 8 (1): 3-18.

Silander, J.A., Jr. 2000. Temperate forests: plant species biodiversity and conservation, In *Encyclopedia of Biodiversity*, S.A. Levin, ed., Academic Press, New York, pp. 5:607-626.

Simonetti, J. A., M. T. K. Arroyo, A. E. Spotorno & E. Lozada (comps.). 1995. *Diversidad Biológica de Chile*, CONICYT, Santiago de Chile.

Tala, C. 1991. Avifauna observada en la Reserva Nacional Río de los Cipreses (VI Región). Boletín Informativo Unión de Ornitólogos de Chile (UNORCH) (Chile) 12: 13-15.

Torres-Mura, J.C. 1994. Estado de conservación de la fauna terrestre de Chile. En: Espinoza G, P Pisani, L Contreras & P Camus (Eds.) Perfil Ambiental de Chile: 367-375. Comisión Nacional del Medio Ambiente, Santiago. 569 pp.

Veblen, T., R.S. Hill & J. Read (eds.). 1996. *The Ecology and Biogeography of Nothofagus Forests,* Yale University Press.

Veblen, T.T., F.M. Schlegel & J.V. Oltremari. 1983. Temperate broad-leaved evergreen forests of South America. En Ovington, J.D. (ed.), *Temperate Broad-leaved Evergreen Forests*. Ecosystems of the World Volume 10. Elsevier, Amsterdam, pp. 5-31.

Villagrán, C. & L. F. Hinojosa. 1997. Historia de los bosques del sur de Sudamérica, II: análisis fitogeográfico. *Revista Chilena de Historia Natural* 70:24 1-267.

Vuilleumier, F. 1985. Forest birds of Patagonia: ecological geography, speciation, endemism and faunal history. En: Buckley PA, MS Foster, ES Morton, RS Riedely & FG Buckley (eds) Neotropical Ornithology. Ornithological Monographs 36: 255-304.

Wilcox, K. 1995. *Chile's Native Forests: A Conservation Legacy.* Ancient Forest International, Redway, CA.

Ornithological references

Araya, B., G. Millie & M. Bernal. 1986. *Guía de Campo de las Aves de Chile*. Editorial Universitaria, Santiago, Chile.

Araya, B. & S. Chester. 1993. *The Birds of Chile*. Latour, Santiago, Chile.

Clark, R. 1986. *Aves de Tierra del Fuego y Cabo de Hornos. Guía de Campo*. L.O.L.A., Buenos Aires, Argentina.

Couve, E. & C. Vidal. 2000. *Aves del Canal de Beagle y Cabo de Hornos*. Fantástico Sur, Punta Arenas, Chile.

Couve, E. & C. Vidal. 2003. *Birds of Patagonia, Tierra del Fuego & Antarctic Peninsula*. Editorial Fantástico Sur Birding Ltda., Punta Arenas, Chile.

De la Peña, M. & M. Rumboll. 1998. *Birds of Southern South America and Antarctica*. HarperCollins Publishers, London.

De Schauensee, R. 1970. *A Guide to the Birds of South America*. Academy of Natural Sciences of Philadelphia by Livingston Pub. Co., Wynnewood, Pa.

Egli, G. & J. Aguirre. 2000. *Aves de Santiago*. UNORCH, Z&D Servicios Gráficos, Chile.

Fjeldså, J. & N. Krabbe. 1990. *Birds of the High Andes*. Zoological Museum and Apollo Books, Copenhagen and Stenstrup, Denmark.

Goodall, J.D., A.W. Johnson & R.A. Philippi. 1946. *Las Aves de Chile*. Volumen I. Establecimientos Gráficos Platt S.A., Buenos Aires, Argentina.

Goodall, J.D., A.W. Johnson & R.A. Philippi. 1951. *Las Aves de Chile*. Volumen II. Establecimientos Gráficos Platt S.A., Buenos Aires, Argentina.

Hellmayr, C.E. 1932. Birds of Chile. *Field Museum Natural History Publications 308 (Zoological Series)* 19: 1-472.

Hoffmann, A. & I. Lazo. 2000. *Aves de Chile. Un Libro También para Niños*. Universidad Andrés Bello, Ril Editores, Santiago.

Housse, R.E. 1945. *Las Aves de Chile en su Clasificación Moderna. Su Vida y Costumbres*. Ediciones Universidad de Chile, Santiago.

Jaramillo, A. 2005. *Aves de Chile*. Lynx, Barcelona, España.

Martínez, D. & G. González. 2004. *Las Aves de Chile. Nueva Guía de Campo*. Ediciones del Naturalista, Santiago, Chile.

Moreira, A. 1999. *Guía de Campo. Caleu y Cerro El Roble*. Asociación de Comuneros La Capilla de Caleu, Chile.

Narosky, T. & A. Bosso. 1995. *Manual del Observador de Aves*. Editorial Albatros Saci, Buenos Aires.

Narosky, T. & D. Yzurieta. 1987. *Guía para la Identificación de las Aves de Argentina y Uruguay*. Vazquez Mazzini Editores, Buenos Aires.

Narosky, T. & M. Babarskas. 2000. *Guía de Aves de Patagonia y Tierra del Fuego*. Vázquez Mazzini Editores, Buenos Aires, Argentina.

Rozzi, R., D. Martínez, M.F. Willson & C. Sabag. 1996. Avifauna de los Bosques Templados
    de Sudamérica. In *Ecología de los Bosques Nativos de Chile* (J.J. Armesto, C. Villagrán & M.T. Kalin,
    eds.), pp: 135-152. Editorial Universitaria, Santiago, Chile.
Venegas, C. 1994. *Aves de Magallanes*. Ediciones de la Universidad de Magallanes,
    Punta Arenas, Chile.
Vuilleumier, F. 1985. Forest birds of Patagonia: ecological geography, speciation, endemism, and faunal
    history. *Neotropical Ornithology, Ornithological Monographs* 36: 255-304.
Willson, M.F., T.L. de Santo, C. Sabag & J.J. Armesto. 1994. Avian communities of fragmented
    south-temperate rainforests in Chile. *Conservation Biology* 8:508-520.

References used in the preparation of bird species distribution range maps

Allen, T.B. (ed.). 1983. *Field Guide to the Birds of North America*. National Geographic Society, Washington.
Barros, V.A. 1976. Nuevas aves observadas en las Islas Picton, Nueva, Lennox y
    Navarino oriental. *Anales del Instituto de la Patagonia* 7: 190-193.
Barros, A. 1971. Aves observadas en las Islas Picton, Nueva, Lennox y Navarino Oriental. *Anales del Instituto
    de la Patagonia* 2: 166-180.
Blake, E.R. 1977. *Manual of Neotropical Birds*. University of Chicago Press, Chicago.
Butler, T.Y. 1979. *The Birds of Ecuador and the Galapagos Archipelago: A Checklist of all the Birds Knownin
    Ecuador and the Galapagos Archipelago and a Guide to Help Locate and See Them*. Ramphastos
    Agency, Portsmouth, N.H.
Davis L.I. 1972. *A Field Guide to the Birds of Mexico and Central America*. University of Texas Press, Austin.
De La Peña, M. 1988. *Guía de Aves Argentinas. Columbiformes a Piciformes*. Facultad de Agronomía y
    Veterinaria, U.N.L.
De la Peña, R. & M. Rumboll. 1998. *Birds of Southern South America and Antarctica*. Collins Illustrated
    Checklist. HarperCollins*Publishers*, Rotolito Lombarda, Italy.
De Schauensee, R.M. & W.H. Phelps, Jr. 1978. *A Guide to the Birds of Venezuela*. Princeton University Press,
    Princeton, N.J.
De Schauensee, R.M. 1964. *The Birds of Colombia, and Adjacent Areas of South and Central America*.
    Livingston Pub. Co. Narberth, Pa.
De Schauensee, R.M. 1966. *The Species of Birds of South America and their Distribution*. Academy of Natural
    Sciences; dist. by Livingston, Narberth, Pa., Philadelphia.
del Hoyo, J., A. Elliott & J. Sargatal (eds). 1992. *Handbook of the Birds of the World. Volume 1: Ostrichs to
    Ducks*. Lynx Editions, Barcelona.
del Hoyo, J., A. Elliott & J. Sargatal (eds.). 1994. *Handbook of the Birds of the World. Volume 2: New World
    Vultures to Guineafowls*. Lynx Editions, Barcelona.
del Hoyo, J., A. Elliott & J. Sargatal (eds.). 1996. *Handbook of the Birds of the World. Volume 3: Hoatzin to
    Auks*. Lynx Edicions, Barcelona.
del Hoyo, J., A. Elliott & J. Sargatal (eds.). 1997. *Handbook of the Birds of the World. Volume 4: Sandgrouse to
    Cukoos*. Lynx Editions, Barcelona.
del Hoyo, J., A. Elliott & J. Sargatal (eds.). 1999. *Handbook of the Birds of the World. Volume 5: Barn Owls to
    Hummingbirds*. Lynx Editions, Barcelona.
del Hoyo, J., A. Elliott & J. Sargatal (eds.). 2001. Handbook of the Birds of the World. Volume 6: Mousebirds to
    Hornbills. Lynx Editions, Barcelona.
Dunning, J.S. 1982. *South American Land Birds*. Harrowood Book, Newtown Square, Pennsylvania.
Elphick, C., J.B. Dunning Jr. & D. A. Sibley (eds.). 2001. *The Sibley Guide to Bird Life & Behavior*.
    National Audubon Society, Alfred A. Knopf, New York.

Garrido, O.H. & F. García-Montaña. 1975. *Catálogo de las Aves de Cuba.* La Habana, Cuba.

Heintzelman, D.S. 1979. *Hawks and Owls of North America: A Complete Guide to North American Birds of Prey.* Universe Books, N.Y.

Johnsgard, P.A. 1990. *Hawks, Eagles and Falcons of North America.* The Smithsonian Institute Press, Washington D.C.

Land, H. 1970. *Birds of Guatemala.* Published for the International Committee for Bird Preservation, Pan-American Section by Livingston Pub. Co., Wynnewood, Pa.

Monroe, B.L. Jr. 1968. *A Distributional Survey of the Birds of Honduras.* Ornithological Monographs N°7. American Ornithologists' Union, N.Y.

Narosky, T. & M. Babarskas. 2000. *Aves de la Patagonia. Guía para su Reconocimiento.* Vásquez Mazzini Editores, Buenos Aires.

Olrog, C.C. 1963. *Lista y Distribución de las Aves Argentinas.* Tucumán, Argentina.

Parker, T.A. III, D.F. Stotz & J.W. Fitzpatrick. 1996. *Neotropical Birds: Ecology and Conservation.* University of Chicago Press, Chicago.

Paynter, R.A. & M.A. Traylor, Jr. 1977. *Ornithological Gazetteer of Ecuador.* Bird Dept., Museum of Comparative Zoology, Harvard University, Cambridge, Mass.

Paynter, R.A. & M.A. Traylor, Jr. 1981. *Ornithological Gazetteer of Colombia.* Bird Dept., Museum of Comparative Zoology, Harvard University, Cambridge, Mass.

Paynter, R.A. Jr. 1982. *Ornithological Gazetteer of Venezuela.* Bird Dept., Museum of Comparative Zoology, Harvard University, Cambridge, Mass.

Paynter, R.A. Jr. 1989. *Ornithological Gazetteer of Paraguay.* Bird Dept., Museum of Comparative Zoology, Harvard University, Cambridge, Mass.

Paynter, R.A., Jr., M.A. Traylor, Jr. & B. Winter. 1975. *Ornithological Gazetteer of Bolivia.* Bird Dept., Museum of Comparative Zoology, Harvard University, Cambridge, Mass.

Perrins, C.M. 1990. *The Illustrated Encyclopedia of Birds: The Definitive Reference to Birds of the World.* Prentice Hall Editions, N.Y.

Pinto, O.M.O. 1960. *Beija-flores do Brasil.* Livraria São José, Rio de Janeiro.

Rand, D.M. & R.A. Paynter, Jr. 1981. *Ornithological Gazetteer of Uruguay.* Bird Dept., Museum of Comparative Zoology, Harvard University, Cambridge, Mass.

Remsen, Jr, J.V. & M.A. Traylor, Jr. 1989. *An Annotated List of the Birds of Bolivia.* Buteo Books, Vermillion, South Dakota.

Ridgely, R.S. & G. Tudor. 1989. *The Birds of South America.* Volume 1, The Oscine Passerines. University of Texas Press, Austin.

Ridgely, R.S. & G. Tudor. 1994. *The Birds of South America.* Volume 2. University of Texas Press, Austin, Texas.

Ridgely, R.S. & J.A. Gwynne, Jr. 1989. A Guide to the Birds of Panama, with Costa Rica, Nicaragua, and Honduras. 2nd ed. Princeton University Press, Princeton, N.J.

Robbins, C.S., B. Bruun & H.S. Zim. 1983. *Birds of North America: A Guide to Field Identification.* Golden Press, N.Y.

Sibley, C.G. & B.L. Monroe, Jr. 1990. *Distribution and Taxonomy of Birds of the World.* Yale University Press, N.H.

Snyder, D.E. 1966. *The Birds of Guyana (formerly British Guinna); A Checklist of 720 Species, with Brief Descriptions, Voice and Distribution.* Peabody Museum, Salem.

Stephens, L. & M.A. Traylor, Jr. 1985. *Ornithological Gazetteer of the Guianas.* Bird Dept., Museum of Comparative Zoology, Harvard University, Cambridge, Mass.

Stiles, F.G. & A.F. Skutch. 1989. *A Guide to the Birds of Costa Rica.* Comstock, Ithaca.

Texera, W. 1972. Distribución y diversidad de mamíferos y aves en la Provincia de Magallanes. Análisis preliminar de la diversidad ecológica y variación taxonómica. Anales del Instituto de la Patagonia (Chile) 3: 171-200.

Texera, W. 1973. Distribución y diversidad de mamíferos y aves en la Provincia de Magallanes. II: Algunas notas ecológicas sobre los canales patagónicos. Anales del Instituto de la Patagonia (Chile) 4(1-3): 291-305.

Venegas, C. 1981. Aves de las Islas Wollaston y Bayly, Archipiélago del Cabo de Hornos. *Anales del Instituto de la Patagonia* 12: 213-219.

Venegas, C. 1991. Ensambles avifaunísticos estivales del Archipiélago Cabo de Hornos. *Anales del Instituto de la Patagonia* 20: 69-81.

Vigil, C. 1973. *Aves Argentinas y Sudamericanas*. Editorial Atlántida, Buenos Aires.

Woods, R.W. 1975. *The Birds of the Falkland Islands*. Anthony Nelson, Oswestry.

Zalles, J.I. & K.L. Bildstein (eds). 2000. *Raptor Watch: A Global Directory of Raptor Migration Sites*, Birdlife International, Cambridge.

Webpages

Aves de Chile:

http://www.avesdechile.cl/

UNEP-WCMC Species Database:

http://www.unep-wcmc.org/isdb/Taxonomy/index.cfm?displaylanguage=ENG

The IUCN Red List of Threatened Species:

http://www.iucnredlist.org/apps/redlist/details/141778/0

Birdlife International:

http://www.birdlife.org/

223

PARTICIPANTS

Lorenzo Aillapan
Poet, Mapuche Bird Man
Academia Mapuche Püllümapukimunweftuy
Puerto Saavedra, IX Región, Chile

Christopher B. Anderson
Ecologist, Ph.D.
Sub-Antarctic Biocultural Conservation Program
University of North Texas - OSARA
Omora Ethnobotanical Park - Universidad de Magallanes, Chile
E-mail: Christopher.Anderson@unt.edu

Uta Berghöefer
Geographer, Ph.D.(c)
Helmholtz Centre for Environmental Research, Leipzig,
Germany
E-mail: Uta.Berghoefer@ufz.de

Alejandra Calcutta
Graphic Designer
Studio Ochenta, Punta Arenas, Chile
E-mail: studio80@tie.cl

Úrsula Calderón
Artisan
Comunidad Indígena Yagán de Bahía Mejillones
She lived on Navarino Island, Chile, until January 2003.
Today she rests in the Cemetery of Mejillones Bay.

Cristina Calderón
Artisan
Comunidad Indígena Yagán de Bahía Mejillones
Villa Ukika, Puerto Williams, Isla Navarino, Chile.

George A. Clark, Jr.
Ornithologist, Ph.D.
Professor Emeritus, University of Connecticut
E-mail: george.clark@myfairpoint.net

Guillermo Egli
Teacher and Ornithologist
AVESCHILE (ex-UNORCH)
gegli@ctcinternet.cl

Luis Gómez
School Teacher
Comunidad Indígena Yagán de Bahía Mejillones, Villa
Ukika, Isla Navarino Chile

María Rosa Gallardo
Chemistry Engineer
GIS Laboratory, Universidad de Magallanes, Chile
E-mail: maria.gallardo@umag.cl

Nolberto González
Music Teacher
Estudios Button Records, Santiago, Chile.
E-mail: buttonrecords@hotmail.com

Kelli Moses
Biology student
University of North Texas
E-mail: kelli.moses@gmail.com

Francisca Massardo
Plant Physiologist, Ph.D.
Omora Ethnobotanical Park
Universidad de Magallanes – Institute of Ecology and
Biodiversity, Chile
Puerto Williams, Chile
E-mail: massardorozzi@yahoo.com;
francisca.massardo@umag.cl

Kurt Heidinger
Writer, Ph.D.
Omora Ethnobotanical Park
E-mail: kurtheidinger@yahoo.com

Steven McGehee
Field Ornithologist, M.S.
E-mail: whitethroatedcaracara@yahoo.com

Lorena Peñaranda
Lawyer
Viña del Mar, Chile

José C. Pizarro
Veterinarian, M.S.(c)
Universidad de Magallanes
E-mail: jcpizarrop@gmail.com

Eduardo Ramilo
Ornithologist
Delegación Regional Patagonia, Administración de Parques
Nacionales
Casilla 380, San Carlos de Bariloche, Argentina
nahuelhuapi@apn.gov.ar

Ricardo Rozzi
Philosopher, M.A. & Ecologist, Ph.D.
Sub-Antarctic Biocultural Conservation Program
University of North Texas
Omora Ethnobotanical Park - Universidad de Magallanes,
Chile
E-mail: ricardo.rozzi@unt.edu

John Schwenk
Recording Engineer
E-mail: johnschwenk@juno.com

Paola Vezzani
Visual Artist
Omora Ethnobotanical Park
Universidad de Magallanes
paola.vezzani@gmail.com

Oliver Vogel
Photographer
Puerto Williams, Chile
E-mail: foto-olv@web.de

Cristina Zárraga
Poet and Writer
Comunidad Indígena Yagán de Bahía Mejillones
Puerto Williams, Chile
E-mail: hannuja@yahoo.com

IMAGE CREDITS

Augustin Berghöefer   Page: 20 (recording on Navarino Island).

Eduardo Camelio   Page: 77 (*Glaucidium nanum*).

Enrique Couve   Pages: 134 (*Picoides lignarius*), 125 (*Patagioenas araucana*), 153 (*Colorhamphus parvirostris*), 178 (*Sturnella loyca*).

Iván Díaz   Pages: 87 (*Glaucidium nanum*), 164 (*Curaeus curaeus*).

Emilio García de la Huerta Page: 141 (*Carduelis barbata*).

Gonzalo E. González   Page: 174 (*Diuca diuca*).

Vicente González   Pages: 49 (*Campephilus magellanicus*, female), 56 (*Pteroptochos tarnii*), 66 (*Aphrastura spinicauda*), 70 (*Enicognathus ferrugineus*), 82 (*Bubo magellanicus*), 88 (*Glaucidium nanum*), 107 and 108 (*Cinclodes patagonicus*), 10 (*Tachycineta meyeni*), 123 (*Xolmis pyrope*), 123 and 130 (*Colaptes pitius*), 132 (*Colaptes pitius*), 138 (*Elaenia* albiceps), 144 (*Phrygilus patagonicus*), 151 (*Troglodytes aedon*), 156 y 157 (*Anairetes parulus*), 176 (*Sturnella loyca*), 183, 201 and 202 (*Vultur gryphus*), 188 (*Caracara plancus*), 191 (*Falco sparverius*), 194 (*Geranoaetus melanoleucus*).

José Tomás Ibarra   Pages: 47 (*Aphrastura spinicauda*), 74 (*Sylviorthorhynchus desmursii*), 77 (*Parabuteo unicinctus*), 78 (*Strix rufipes*), 105 (*Cinclodes fuscus*), 113 (*Pygochelidon cyanoleuca*), 115 and 116 (*Theristicus melanopis*), 134 and 135 (*Picoides lignarius*), 176 (*Sturnella loyca*), 189 (*Caracara plancus*).

Jaime Jiménez   Pages: 124 (*Patagioenas araucana*), 183 (*Parabuteo unicinctus*).

Francisca Massardo   Page: 26 (Valdivian rainforests).

Rodrigo Molina   Page: 184 (*Milvago chimango eating*).

Steve Morello   Pages: 47 and 57 (*Pteroptochos tarnii*), 47 and 59 (*Scelorchilus rubecula*), 52 (*Scytalopus magellanicus*), 139 (*Elaenia* albiceps), 150 (*Troglodytes aedon*), 170 (*Sephanoides sephaniodes*), 180 (*Phytotoma rara*), 209 (*Coragyps atratus*).

Joan Morrison   Pages: 183 (*Caracara plancus*), 184 (*Milvago chimango*).

Eduard Müller   Pages: 118 and 119 (*Vanellus chilensis*), 162 (*Xolmis pyrope*).

Omar Ohrens   Page: 90 (*Accipiter bicolor chilensis*).

Carlos Olavarría   Pages: 62 and 64 (*Pygarrhichas albogularis*).

Eduardo Pavéz   Pages: 85 (*Tyto alba*), 196 (*Buteo polyosoma*), 199 (*Parabuteo unicinctus*), 201 (*Vultur gryphus*).

Jordi Plana   Pages: 47 and 48 (*Campephilus magellanicus*, male), 92 (wetland in the Navarino Island), 93 and 98 (*Gallinago paraguaiae*).

Eduardo Ramilo
Pages: 77 (*Bubo magellanicus*), 103 (*Pardirallus sanguinolentus*), 115 (*Theristicus melanopis in group*), 120 (*Vanellus chilensis*), 141 (*Carduelis barbata*), 165 (*Curaeus curaeus*), 185 (*Milvago chimango*), 193 (*Geranoaetus melanoleucus*), 197 (*Buteo polyosoma*), 294 (*Vultur gryphus*).

Marie-Louise Roux
Page: 94 (*Ceryle torquata*).

Ricardo Rozzi
Pages: 16 (workshop in Cucao), 18 (recording group in Mejillones), 19 (WHUS Studio), 25 (*Araucaria araucana*), 26 (Valdivian understory), 27 (Sub-Antarctic forests), 38 (Cristina and Ursula Calderon), 40 (Aillapan at Omora Park), 68 (*Aphrastura spinicauda*), 211 (grandmothers and grandsons at the Omora Park).

Michel Sallaberry
Pages: 71 (*Enicognathus ferrugineus*), 93 and 102 (*Pardirallus sanguinolentus*), 128 (*Zenaida auriculata*), 206 and 207 (*Cathartes aura*).

John Schwenk
Page: 17 (recording with Ursula Calderon).

Margaret Sherriffs
Page: 212 (children and teachers coming in the Omora Park).

Ricardo Stanoss
Pages: 35 (Sub-Antarctic forest at Wulaia Bay, Navarino Island), 95 and 96 (*Ceryle torquata*).

Juan Carlos Torres-Mura
Pages: 80 (*Strix rufipes*), 99 and 100 (*Gallinago paraguaiae*), 111 (*Tachycineta meyeni*), 131 (*Colaptes pitius*), 137 (*Elaenia* albiceps), 147 (*Turdus falcklandii*), 154 (*Colorhamphus parvirostris*), 161 (*Xolmis pyrope*), 172 (*Patagona gigas*), 187 (*Caracara plancus*).

Steff van Dongen
Pages: 72 (*Enicognathus ferrugineus*), 145 (*Phrygilus patagonicus*).

Paola Vezzani
Pages: 46 (forest interior), 76 (dusk at the Strait of Magellan), 67 (*Aphrastura spinicauda*), 93 (*Cinclodes fuscus*) 122 (sub-Antarctic forest ecosystems, Darwin Cordillera, Cape Horn Biosphere Reserve), 127 (*Zenaida auriculata*), 159 (*Zonotrichia capensis*), 182 (landscape of the Fuegia Island).

Luc Viatour
Page: 84 (*Tyto alba*), www.lucnix.be

Oliver Vogel
Pages: 17 (recording with Aillapan at Omora Park), 20 (Button Records Studio), 27 (Decidous Beech forest), 39 (Cristina Calderón), 63 (*Pygarrhichas albogularis*).

Norm Wickett
Page: 79 (*Strix rufipes*).

Günter Ziesler
Pages: 123, 167 and 170 (*Sephanoides sephaniodes*, nesting and flying), 171 (*Patagona gigas*).

Drawings
Jay Barry (caricatures of Humberto Maturana , Charles Darwin and hummingbird in pages 15, 30 and 211). Daniel Martínez (Rufous Collared Sparrow and Omora in page 36).

Figures
Francisca Massardo (Figures 16, 18A , 18B, 212);

Satellite image and maps María Rosa Gallardo

INDEXES

YAHGAN BIRD NAMES

MAPUDUNGUN BIRD NAMES

SPANISH BIRD NAMES

Águila	193	Golondrina de dorso negro	113
Aguilucho	196	Huet Huet	57
Bandurria	115	Jilguero	141
Becasina	99	Jote de cabeza colorada	206
Cachaña	71	Jote de cabeza negra	209
Cachudito	157	Lechuza blanca	84
Carpinterito	134	Loica	177
Carpintero negro	49	Martín pescador	95
Cernícalo	191	Peuco	199
Colilarga	74	Peuquito	90
Comesebo grande	63	Picaflor chico	167
Cometocino	145	Picaflor gigante	171
Concón	79	Pidén	102
Cóndor	201	Pitío	131
Chercán	150	Queltehue	119
Chincol	159	Rara	180
Chucao	59	Rayadito	67
Chuncho	87	Tiuque	184
Churrete	107	Torcaza	125
Churrete acanelado	105	Tordo	165
Churrín	52	Tórtola	127
Diuca	174	Traro	187
Diucón	161	Tucúquere	82
Fío-fío	137	Viudita	153
Golondrina chilena	110	Zorzal	147

ENGLISH BIRD NAMES

SCIENTIFIC BIRD NAMES

RECORDING INDEX
CD I English

	Track	Name	Time
Introduction	1	Trutruka song (Lorenzo Aillapan)	1:06:51
FOREST INTERIOR	2	Magellanic Woodpecker	0:40:57
	3	Magellanic Tapaculo	0:22:36
	4	Black-Throated Huet-Huet	0:32:65
	5	Chucao Tapaculo	0:31:14
	6	White-Throated Treerunner	0:24:40
	7	Thorn-Tailed Rayadito	0:25:40
	8	Austral Parakeet	0:28:57
	9	Desmur´s Wiretail	0:50:67
OWLS	10	Rufous-Legged Owl	0:43:14
	11	Austral Great Horned Owl	0:31:21
	12	Barn Owl	0:27:52
	13	Austral Pygmy Owl	0:28:16
	14	Bicolored Hawk	0:17:52
WETLANDS	15	Ringed Kingfisher	0:31:60
	16	Common Snipe	0:39:17
	17	Plumbeous Rail	0:29:27
	18	Bar-Winged Cinclodes	0:25:18
	19	Dark-Bellied Cinclodes	0:24:02
	20	Chilean Swallow	0:31:26
	21	Blue-and-White Swallow	0:25:24
	22	Buff-Necked Ibis	0:33:21
	23	Southern Lapwing	0:23:18
FOREST MARGINS	24	Chilean Pigeon	0:24:74
	25	Eared Dove	0:31:08
	26	Chilean Flicker	0:20:35
	27	Striped Woodpecker	0:29:28
	28	White-Crested Elaenia	0:25:74
	29	Black-Chinned Siskin	0:24:68
	30	Patagonian Sierrafinch	0:29:02
	31	Austral Thrush	0:26:34
	32	House Wren	0:30:14
	33	Patagonian Tyrant	0:21:39
	34	Tufted Tit-Tyrant	0:30:09
	35	Rufous-Collared Sparrow	0:39:12
	36	Fire-Eyed Diucon	0:35:61
	37	Austral Blackbird	0:35:07
	38	Green-Backed Firecrown	0:19:11
	39	Giant Hummingbird	0:20:06
	40	Common Diuca Finch	0:31:18
	41	long-Lailed Meadowlark	0:27:28
	42	Rufous-tailed Plantcutter	0:30:19
RAPTOR	43	Chimango Caracara	0:30:68
	44	Southern Caracara	0:27:17
	45	American Kestrel	0:22:10
	46	Black-chested Buzzard-eagle	0:22:22
	47	Red-backed Hawk	0:24:48
	48	Bay-Winged Hawk	0:20:68
	49	Andean Condor	0:23:33
	50	Turkey Vulture	0:26:74
	51	Black Vulture	0:29:17

BIRD NAMES

MAPUCHE STORIES

52	Chucao Tapaculo	2:50:15
53	Rufous-Legged owl	1:53:06
54	Ringed Kingfisher	1:21:03
55	Plumbeous Rail	1:54:16
56	Swallow	1:49:55
57	White-crested Elaenia	2:44:42
58	Chilean Flicker	1:29:31
59	Striped Woodpecker	1:33:01
60	Eared Dove	3:03:12
61	Patagonian Tyrant	2:28:26
62	Long-tailed Meadowlark	2:09:36
63	Southern Lapwing	2:46:69
64	Chimango Caracara	1:09:38
65	Southern Caracara	1:15:70
66	Red-backed Hawk	1:47:55
67	Andean condor	2:48:22

Total **58:01**

CD II

Track	Name	Time
1	*Omora* Green-Backed Hummingbird	6:58:69
2	Fire-Eyed Diucon	1:58:07
3	Buff-Necked Ibis	4:38:66
4	Common Snipe	2:27:01
5	Black-Chinned Siskin	5:05:50
6	Barn Owl	2:17:13
7	Thorn-Tailed Rayadito	1:25:58
8	Magellanic Tapaculo	4:47:49
9	White-Throated Treerunner	2:01:11
10	Magellanic Woodpecker	4:25:41
11	Austral Trush	4:03:51
12	Gli Uccellini (composer Andrés Alcalde, interpreter Guillermo Lavado)	1:52:56

Total **42:05:00**

233

The Millennium Science Initiative Program (MSI) is an original model in the developing world whose main objective is to promote the advancement of cutting edge scientific and technological research in Chile. This is done through Centers of Excellence in scientific research in the fields of Natural and Exact Sciences and in Social Sciences, as a relevant actor in the National System of Science, Technology and Innovation. www.iniciativamilenio.cl

This book is a revised and amplified edition of the "Multi-ethnic bird guide of the austral temperate forests of South America," published by Fantástico Sur – Birding & Nature, in 2003.

Cover: Branches of High Deciduous Beech *(Nothofagus pumilio)* with photographs of Úrsula Calderón and Magellanic Woodpecker (top left), Cristina Calderón and Ringed Kingfisher (top right), Lorenzo Aillapan and Red-Backed Hawk, and Ricardo Rozzi and Austral Pygmy Owl. Design by Paola Vezzani & Ricardo Rozzi. Photographs by John Schwenk (Úrsula Calderón and Ricardo Rozzi), Paola Vezzani (branches of High Deciduous Beech or "Lenga" tree, and coastal landscape at the Beagle Channel, Navarino Island), Oliver Vogel (Cristina Calderón and Lorenzo Aillapan), Steve Morello (Austral Pygmy Owl, *Glaucidium nanum*), Eduardo Pavéz (Red-backed Hawk, *Buteo polyosoma*), Jordi Plana (Magellanic Woodpecker, *Campephilus magellanicus*), and Ricardo Stanoss (Ringed Kingfisher, *Ceryle torquata*).

Graphic design and layout of the book by Paola Vezzani and Alejandra Calcutta.

Full citation of the book: Rozzi, R., F. Massardo, C. Anderson, S. McGehee, G. Clark, G. Egli, E. Ramilo, U. Calderón, C. Calderón, L. Aillapan, & C. Zárraga. 2010. Multi-ethnic Bird Guide of the Sub-Antarctic Forests of South America. University of North of Texas Press - Ediciones Universidad de Magallanes, Denton, Texas, and Punta Arenas, Chile.

SALVIAT IMPRESORES S.A.
Seminario 739, Santiago, 56(2) 3418410, esalviat@salviatimpresores.cl
Printed in Santiago - Chile